T0230695

Constructability

Constructability
A Tool for Project Management

Sharmin Khan

CRC Press
Taylor & Francis Group
Boca Raton London New York

CRC Press is an imprint of the
Taylor & Francis Group, an **informa** business

CRC Press
Taylor & Francis Group
6000 Broken Sound Parkway NW, Suite 300
Boca Raton, FL 33487-2742

© 2019 by Taylor & Francis Group, LLC
CRC Press is an imprint of Taylor & Francis Group, an Informa business

No claim to original U.S. Government works
Printed on acid-free paper

International Standard Book Number-13: 978-1-138-58675-8 (Hardback)

Library of Congress Cataloging-in-Publication Data

Names: Khan, Sharmin, author.
Title: Constructability : a tool for project management / Sharmin Khan.
Description: Boca Raton : Taylor & Francis, a CRC title, part of the Taylor & Francis imprint, a member of the Taylor & Francis Group, the academic division of T&F Informa, plc, 2019. | Includes bibliographical references and index.
Identifiers: LCCN 2018020503 | ISBN 9781138586758 (hardback : acid-free paper) | ISBN 9780429504426 (ebook)
Subjects: LCSH: Building—Superintendence. | Construction projects—Management.
Classification: LCC TH438 .K448 2018 | DDC 690.068/4—dc23
LC record available at https://lccn.loc.gov/2018020503

Visit the Taylor & Francis Web site at
http://www.taylorandfrancis.com

and the CRC Press Web site at
http://www.crcpress.com

Contents

Preface

This book was motivated by the desire to strengthen and promote the practices of constructability. My first encounter with the idea of constructability was accidental, during my master's program at IIT Delhi, New Delhi, India, where I was allotted a project for research on constructability. The more I read, the more curious I was, and the eagerness to enquire into the relationship between constructability and sustainability paved way to my doctoral research. The investigation was initiated to enquire, "Is it possible to enhance sustainability in building projects by adopting constructability practices?" which seemed to be exciting. This book presents a review study of other researchers' work who have been actively involved in the studies focused on constructability practices worldwide together with the study conducted by me in the Indian construction industry.

Project management in the construction industry is all about bringing together people of varying specializations and coordinating them to accomplish the task. The idea of presenting this work emerged due to the concern for the lack of coordination among team members, which ultimately leads to loss of economy. The dynamic changes in the trends of handling construction projects was the driving force behind the evolution of the concept of constructability. This realization during the last couple of decades is the key to various research projects on constructability.

This book focuses on the need to introduce and implement the idea of constructability in construction projects. The contents are broadly divided into three sections. Part I presents the research performed worldwide on different issues of constructability. The benefits are analyzed, and the barriers to the implementation of constructability are presented with a hope to break them off. Some of the researchers worldwide have been engaged in conducting surveys from time to time to analyze the scenario in the construction industry regarding the constructability practices followed. A comprehensive study of these research projects is also presented in this section.

Part II presents ten case studies performed in the Indian construction industry. The relationship between constructability and sustainability is examined in this section. In case such a relationship exists, the problem is half solved because the stakeholders in the construction industry can contribute to sustainable development by improvisation in their management practices. This section also identifies the interrelationship among design and construction activities (based on constructability). The establishment of such relationship shall enforce the idea of integration and coordination among team members.

Part III brings together the recommendations extracted from discussions with the practitioners during research work, presented in Part II. Few models focusing on the leadership of a project are also discussed. This section concludes with checklists for circumspection during different stages of building a project.

This piece of work is based on my doctoral research "A study of design and construction practices of sustainable architecture in India based on the concept of constructability" submitted at AMU, ALigarh and is presented for readers as a comprehensive summary of research projects conducted so far. At the same time it recommends the practices of constructability for promotion of sustainable development.

Sharmin Khan

About the Author

Sharmin Khan is a graduate in architecture from Aligarh Muslim University (AMU), Aligarh, India, and postgraduate in construction engineering and management from IIT Delhi, New Delhi, India. Collectively, she has 14 years of experience teaching in three Indian universities, including AMU, Aligarh. Her research areas of interest are construction management, sustainable architecture, and history. Her interaction with students during the last couple of years encouraged her to publish books on ancient Indian and Indo-Islamic architecture also. She worked on constructability issues during her postgraduation program, which inspired her to conduct her doctoral studies in the same area. She obtained a PhD in architecture from AMU, Aligarh, focusing on the study of design and construction practices of sustainable architecture in India based on the concept of constructability. Her research work is oriented toward the study and correlation of construction management issues with sustainable development. She has presented and published research papers to promote the practices of constructability in the Indian construction industry.

CHAPTER 1

The Foundation of Constructability

1.1 INTRODUCTION

Construction industry is extremely important worldwide as it provides infrastructural facilities for people and it is a huge employment-generating sector. Unfortunately, there are some management issues that lead to delays in projects and hence loss of economy. Therefore, this emanates the need for introduction of constructability in construction industry. This section highlights the importance and need to promote the concept of constructability and its evolution.

1.2 THE CONSTRUCTION INDUSTRY

Construction industry is one of the biggest industries in the world and contributes toward the gross domestic product of the country. The construction industry creates huge infrastructural facilities for the masses to use and enjoy. It generates employment opportunities for the communities. On the other hand, it is the biggest source of creating pollution and exhausting the nonrenewable resources of energy.

The construction industry has major environmental impacts. Redclift (2005) states that each scientific problem resolved by human intervention using fossil fuels and which manufactures materials is conventionally reviewed as a triumph of management and contribution to economic good; however, it is also seen as a future threat to sustainability. In 1990, residential, commercial, and institutional building sectors globally consumed 31% of global energy and emitted 1,900 megatons of carbon. It is expected that by 2050, this share will rise to 38% and emit 3,800 megatons of carbon (Watson et al., 1996).

The consumption of total natural resources is 50%, energy usage is 40%, and water consumption is 16%. The waste production during construction and demolition is more than the volume of household waste (Muller, 2002).

The building sector has largest potential for energy efficiency. The sustainable construction shall focus on three main areas of life cycle processes of building: its construction, utilization, and demolition or rehabilitation. The self-sustained building concept must have a closed cycle of flows of energy, water, and materials (TERI, 2004). The construction industry is essentially a service industry whose responsibility is to convert plans and specifications into finished products: it is exceedingly complex and highly individual in character (Peurifoy and Ledbetter, 1985). The construction industry consumes large amount of energy, water, materials, and land. This contributes to the exhaustion of natural resources and consumption of energy (Poon, 2000; Shen et al., 2000).

Shen et al. (2004) have mentioned various organizations, those which have been working on environment management systems, such as Building Research Establishment Environmental Assessment Methodology in the United Kingdom, the Building Environmental Performance Assessment Criteria in Canada, the Green Building Challenge in the United States, and Hong Kong Building Environment Assessment Method in Hong Kong. The Chartered Institute of Building in 1989 has also identified certain areas for environmental management in construction activities. Some of these areas are identified as efficient use of energy, environment-friendly building materials, control of toxic chemicals, pollution control, recycling, and waste management (CIOB, 1988). The traditional design and construction practices focus on cost performance and quality issues. Sustainable design and construction adds the issues of minimization of resource consumption, environmental degradation, and creation of healthy, built environment as well as ensuring human health and comfort (Sev, 2008).

The client and the contractor may not be interested in the energy-efficient designs, and the reasons could be that the benefits of such designs can be realized only in the long term, whereas the business in construction industry is oriented toward short-term profit making (Tai, 2000). Various construction activities, such as generation of excessive noise, dust, chemical particles, odor, toxic gases, and solid wastes can cause pollution and harm the environment (Shen et al., 2000). A paper by Riley et al. (2003) analyzed that contractors can help in achieving sustainable project objectives by providing conceptual estimating services during preconstruction, sourcing and procuring sustainable materials, managing construction waste, and helping to ensure that indoor air quality requirements are met.

India has a significant role in the global scenario with respect to some of the major issues related to design and construction practices. Reddy (2004) states that Indian construction industry is one of the largest in terms of economic expenditure, volume of raw material and products manufactured, employment generated, environmental impacts, etc. It is estimated that 22% of the greenhouse gas emission is contributed by construction sector in India. The total quantum of construction and demolition waste generated in India is estimated to be 12–14.7 million tonnes per annum (TIFAC, 2000). The waste is generated from construction industries, which include wasted sand, gravel, bitumen, bricks, masonry, and concrete. The present waste handling practices are not sufficient in terms of efficiency and recovery. The waste is generally disposed without segregation (Thomas and Wilson, 2013).

According to Widermuth (2008), there is an urgent need for a government-sponsored 40–50 year holistic infrastructure plan for India to continue to its high growth path toward economic maturity. Currently, the construction industry in India is facing many barriers and challenges such as lack of awareness among project participants, lack of interest from clients, lack of skilled labor, lack of market competition in terms of saving cost through waste minimization and management techniques, lack of proper training and education among clients and contractors through federations and professional institutes in terms of natural policies and regulations implementation, and lack of waste reduction approach by architects during the planning stage.

In 2012, the Supreme Court of India asked state governments to amend the rules to regulate mining of minerals and ensure environmental management. On August 2013, the National Green Tribunal declared that sand mining without environmental clearance is illegal. India needs to introduce some more policies as an emergent need in the construction industry. These include making of Bureau of Indian Standards codes on recycled material, promotion of alternative materials, introduction of tax policies for minimizing waste, and promotion of efficient construction management practices (CSE, 2014).

1.3 THE NEED FOR CONSTRUCTABILITY

Traditionally, the independent designer was under contract to client and owed a duty to client. He was under no legal obligation to introduce good buildability in his design. There is long-established tradition that the builder should do as the designer tells him and that it is not his province to suggest amendments to the designer. This lack of dialog affects the design. Increasing amount of rework is a proof of inefficient and uneconomical construction. It is a result of poor construction techniques or poor construction management. There is an emergent need for the introduction of processes which can integrate the design and construction activities. A joint approach is the need of time so that the two distinct processes of design and construction go hand in hand. The level of tolerance and acceptance of the other professional's qualities and input is a major issue that needs to be addressed.

Also, the increasing levels of competition and the introduction of manufacturing concepts within the industry led to specialization with the passage of time. Such specialization led to the separation of design and construction facilities. The increasing complexity of many projects makes it more and more difficult for the designers to be fully aware of all the implications of their designs on the construction costs. With the problems associated with separated design and construction continuing to grow, the industry began implementing value engineering and construction management services, recognizing the benefits of constructor's involvement during planning and design phases to reduce the project life cycle cost.

Constructability is a construction management approach that links design and construction processes, which have been isolated in the industry in the current scenario. Making use of construction knowledge, from the earliest stages of a project where the ability to influence cost is at greatest, makes sense from both practical and financial viewpoints. Constructability enhances the quality of constructed facility by better communication among project participants such as design, engineering, and construction professionals. Better communication among these participants reduces the project failure and other related problems.

1.4 THE CONCEPT EVOLUTION

The increasing complexity and specialization in the projects has changed the scenario of construction industry today. The traditional system of design–build type of projects is replaced by a system of multiple contracts in the projects. The more the number of participants, the more the management required. The constructability in building projects has been an object of research since the 1970s in the United Kingdom and the United States. This concept is focused on the early involvement of construction knowledge and experience in planning, engineering, and procurement and field operations to achieve the objectives of the project.

During the 1970s, some studies were conducted in the United Kingdom and the United States, which aimed at maximizing the efficiency of construction projects through the concept of constructability. The Business Roundtable published "The Construction Industry Cost Effectiveness Project" in 1983 to motivate the stakeholders, to improve their work methods, and for cost-effectiveness. The Business Roundtable team had representatives from all groups of the construction industry. The summary report, "More Construction for the Money," defined a problem and proposed actions to address them (Business Roundtable, 1982):

Problem: There is a lack of knowledge by owners with respect to opportunities for cost reduction and shortened schedules by integrating advanced construction methods and materials into the planning, design, and engineering phases of the project.

Action by owners individually: Write contracts that give contractors an incentive to mesh engineering and construction expertise with the process called "constructability," which can often save 10–20 times the cost it adds to a project.

Action by owners jointly: Make concerted efforts to help overcome the shortage of experts in "constructability" by helping to develop training materials and encouraging universities and colleges to add this facet of construction management to their undergraduate curricula.

Action by academia: "Constructability" skills need to be added to undergraduate curricula in construction management.

The efforts of the Business Roundtable led way to the formation of the Construction Industry Institute (CII) based at the University of Texas, Austin, Texas. The term "constructability" was collectively used in the U.S. construction industry for the first time (Pocock et al., 2006). This organization comprises research organizations, construction companies, owners, private and academic institutions, etc.

The Construction Industry Research and Information Association (CIRIA) is another nonprofit organization and works for the improvement of industry. CIRIA also laid emphasis on the design process and early involvement of construction expertise. The concept of constructability was very well promoted by the CII. They are leading researcher and formulated guidelines for implementing constructability (Trigunarsyah, 2004; Wong et al., 2006).

CII has defined constructability as "the optimum use of construction knowledge and experience in planning, engineering, procurement and field operations to achieve overall project objectives" (CII, 1986). The ability to influence the cost of project decreases with time; hence, there is maximum scope in the beginning of the project to consider issues that can affect the cost.

Among various other principles, the involvement of construction knowledge in conceptual planning stage is the most important and basic principle. CII Australia proposed 12 principles for execution of the constructability program. These principles are integration, construction knowledge, team skills, corporate objectives, available resources, external factors, programmer, construction methodology, accessibility, specifications, construction innovation, and feedback. In the 1990s, some studies were conducted at Singapore under the first assessment system for buildability of designs, and the results proved that the lack of integration of construction knowledge into the design process resulted in the exceeding budgets and scheduled deadlines of projects (Trigunarsyah, 2004; Wong et al., 2006). The evolution of this concept of constructability was followed by various research projects, which are still going on.

REFERENCES

Business Roundtable. (1982). Construction industry cost effectiveness project report. NY: Business Roundtable.
Central Pollution Control Board (CPCB). (2000). Management of municipal solid wastes. [Report]. New Delhi: MOEF.

CIOB. (1988). Design and build: Code of estimating practice supplement No.2: The Chartered Institute of Building (CIOB).

Construction and Demolition Waste. (2014). New Delhi: CSE. Retrieved from www.cseindia.org/userfiles/Construction-and%20 -demolition-waste.pdf.

Construction. Industry Institute, CII. (1986). Constructability: A primer. [Report]. Austin: University of Texas.

Muller, D. G. (2002). *Sustainable Architecture and Urbanism: Concepts, Technology, Example*. Basel: Birkhuser.

Peurifoy, R. L., & Ledbetter, W. B. (1985). *Construction Planning, Equipment and Methods*. New York: McGraw-Hill.

Pocock, J. B., Kuennen, S. T., Gambatese, J., & Rauschkolb, J. (2006). Constructability state of practice report. *Journal of Construction Engineering and Management, ASCE, 132*(4), 373–383. DOI:10.1061/(ASCE)0733-9364(2006)132:4(373).

Poon, C. S. (2000). Management and recycling of demolition waste in Hong Kong. *Proceedings of the 2nd International Conference on Solid Waste Management* (pp. 433–442), Taipei, Taiwan.

Redclift, M. (2005). Sustainable development. (1987–2005): An oxymoron comes of age. *Sustainable Development. 13*(4), 212–227. DOI:10.1002/sd.281.

Reddy, B. V. V. (2004). Sustainable building technologies. *Current Science, 87*(7), 899–907.

Riley, D., Pexton, K., & Drilling, J. (2003). Defining the role of contractors on green building projects. *Proceedings of CIB 2003 International Conference on Smart and Sustainable Built Environment (SASBE)*, Brisbane, Australia.

Sev, A. (2008). How can the construction industry contribute to sustainable development? A conceptual framework. *Sustainable Development, 17*, 161–173. DOI:10.1002/sd.373.

Shen, L. Y., Bao, Q., & Yip, S. L. (2000). Implementing innovative functions in construction project management towards the mission of sustainable environment. *Proceedings of the Millennium Conference on Construction Project Management: Recent Developments and the Way Forward* (pp. 77–84), Hong Kong.

Shen, L. Y., Linda, F. C. N., Lucille, W., & Paul, F. (2004). Implementing environmental management in construction projects. *NICMAR Journal of Construction Management, XIX*(II), 1–17.

Tai, Y. S. S. (2000). Environment management system and its application to the building industry in Hong Kong. *The Hong Kong Surveyor, 11*(4), 9–23.

TERI. (2004). *Sustainable Building, Design Manual, Vol. 1.* New Delhi: TERI.

Thomas, J., & Wilson, P. M. (2013). Construction waste management in India. *American Journal of Engineering Research, 2,* 6–9.

TIFAC. (2000). Utilization of waste from construction industry, Department of Science and Technology. New Delhi, India. http://tifac.org.in/index.php/8-publication/184-utilisation-of-waste-from-construction-industry.

Trigunarsyah, B. (2004). A review of current practice in constructability improvement: Case studies on construction projects in Indonesia. *Construction Management & Economics, 22*(6), 567–580. DOI:10.1 080/0144619042000202870.

Watson, R. T., Zinyowera, M. C., & Moss, R. H. (Eds). (1996). *Technologies, Policies and Measures for Mitigating Climate.* Geneva: IPCC.

Widermuth, B. (2008). India, a building industry in transition. *Construct, 6,* 21–28.

Wong, F. W. H., Saram, D. D. D., Lam, P. T. I., & Chan, D. W. M. (2006). A compendium of buildability issues from the viewpoints of construction practitioners. *Architectural Science Review, 49*(1), 81–90.

CHAPTER 2

An Overview of Constructability Practices

2.1 INTRODUCTION

Various research projects conducted on constructability are presented in this section. The constructability issues have been identified and discussed here under 16 categories as follows: integration, coordination, bidding process, construction-driven schedule, simplification of design, standardization of elements, prefabrication, accessibility to site, adverse weather conditions, technical specifications, encouragement to innovations, past lessons learned exercise and reviews, availability of resources, recycling, waste management, and application of advance information technology.

2.2 THE RESEARCH ON CONSTRUCTABILITY

Various researchers have focused on the concept of constructability, after its worldwide importance was recognized. Glavinich (1995) describes constructability of a design as "the ease with which the raw materials of the construction process (labour, production, equipment, tools, materials and installed equipment) can be brought together by a builder to complete the project in a timely and economic manner." Fischer and Tatum (1997) have quoted in their paper the definitions of Buildability and Constructability according to the United Kingdom. Buildability is defined as "the extent to which the design of building facilitates ease of construction, subject to overall requirement for the completed building." U.K. definition for Constructability is "it is the extent to which the design of building facilitates ease of construction, subject to the requirements of construction methods." Buildability focusses on design whereas constructability takes into consideration both, the design and management issues. Constructability incorporates project management systems in the construction project and the benefits are perceptible when constructability is introduced at an early stage (Wong et al., 2006).

O'Connor et al. (1987) have presented and analyzed seven concepts for improving constructability during engineering and procurement phase of the project. These concepts are construction-driven schedule, simplified designs, standardization, module engineering, accessibility, adverse weather, and specifications.

In another paper, O'Connor et al. (1988) have stated previously determined concepts related to constructability under the following heads:

Conceptual planning stage

- Constructability programs are made integral part of project execution plans.

- Project planning actively involves construction knowledge and experience.

- The source and qualifications of personnel with construction knowledge and experience varies with different contracting strategies.

- Overall project schedules are construction sensitive.
- Basic design approaches consider major construction methods.

Design and procurement stage

- Site layout promotes efficient construction.
- Design and procurement schedules are construction sensitive.
- Designs are configured to enable efficient construction.
- Design elements are standardized.
- Project constructability is enhanced when construction efficiency is considered in specification development.
- Module/preassembly designs are prepared to facilitate fabrication, transportation, and installation.
- Designs promote construction accessibility of personnel, material, and equipment.
- Designs facilitate construction under adverse weather conditions.

Field operations stage

- Innovative definitive sequencing of field tasks.
- Innovative uses of temporary construction materials/systems.
- Innovative uses of hand tools.
- Innovative uses of construction equipment.
- Constructor optional preassembly.

- Innovative temporary facilities directly supportive of field methods.

- Post-bid constructor preferences related to the layout, design, and selection of permanent materials.

Tatum (1987) investigated 15 projects and identified three key issues during conceptual planning stage: developing the project plan, laying out the site, and selecting major construction methods. These issues were found beneficial in improving constructability. Radtke (1992) outlined his research looking at constructability practices to integrate the construction knowledge into design and planning phases of project. These methodologies may be either formal or informal ways. Formal ways are identified as documentation, tracking through past lessons learned, and team building exercises and the participation of construction personnel in project planning. The informal ways be like design reviews and inclusion of construction coordinators. Nima et al. (2001) have developed 23 constructability philosophy concepts throughout different phases of construction process as Conceptual planning phase, Design and Procurement phase, and Field operations phase.

Pocock et al. (1996) presented that one of the "critical factors" identifying successful projects is "constructability information from and available to the project team in a timely manner." Constructability program implementation has resulted in significant gains in safety performance, schedule and project cost control (Jergeas and Put 2001). Pulaski and Horman (2005) introduced a model CPPMM (Conceptual Product/Process Matrix Model) for organizing constructability information based on timings and levels of detail. They concluded that "the key to accessing constructability is introducing the right information at the right time and in the right level of detail." In another paper, Pulaski et al. (2006) concluded and evaluated four constructability practices that were used to manage sustainability building knowledge at the renovation of Pentagon. These were (1) an integrated project team, (2) physical and computer models, (3) an on-board review process, and (4) a lessons learned workshop.

2.3 THE IMPORTANT PARAMETERS OF CONSTRUCTABILITY

Some of the major issues have been extracted from the research of various authors and organized under various heads for detailed discussion. The issues have been taken which were common to most of the researchers and the viewpoints gathered thereof. These 16 issues have been identified and listed as follows: integration, coordination, bidding process, construction-driven schedule, simplification of design, standardization of elements, prefabrication, accessibility to site, adverse weather conditions, technical specifications, encouragement to innovations, past lessons learned exercise and reviews, availability of resources, recycling, waste management, and application of advance information technology.

2.3.1 Integration

The integration of all the team members at initial stage of design is an important parameter. But in the existing system, the contractors are involved in the project after the design stage is completed, which is not the ideal integration procedure. Several researchers have worked on this issue and highlighted the importance of integration in construction projects at initial stage of design because the probability of cost saving is maximum at this stage.

Itami and Roehl (1987) identified integration as an "invisible asset." O' Connor et al. (1987) writes that the process of schedule development should involve an interdisciplinary team expert and well represented by construction personnel. The experienced construction personnel should be available on a continuing or timely basis so that they can give their input to the design team. Construction expertise can also help in identifying potential areas where standardization can be applied in the design. Timely review of project by construction personnel can also minimize accessibility problems on site and hence improve the working.

The Business Roundtable's Construction Industry Cost effectiveness Project (Business Roundtable, 1982) has laid emphasis on the participation of constructional experts in the conceptual development stage and the planning stage. The results of this involvement may lead to savings in cost of the project.

Nam and Tatum (1992) highlighted the importance of interorganizational relations as a means of achieving integration. The construction teams are temporary organizations, which come together for a specific purpose of building a facility. Such an organization is for short duration but depends on long-term relations between the owners, engineers, contractors, and suppliers. This relationship is based on trust, reputation, and single-goal achievement concept. O'Connor and Miller (1994) identified certain barriers that do not allow early involvement of contractor, which can be stated as contracting practice, teamwork, and culture. There is a lot of resistance on account of the prevailing culture of adoption of contractor after the design has been finalized.

Glavinich (1995) discusses that construction manager should be involved as soon as possible in the project, so that he can bring advantage to the project through his expertise during early stage of design. Pocock et al. (1996) found that "it is generally accepted that project performance can be enhanced when interaction occurs on a regular basis, beginning at an early stage in a project, in an open and trusting environment." Kichuk and Wiesner (1997) suggested that the process of selection of the firm's professional composition should take place before the beginning of the project. This increases the probability of success of the team. Uhlik and Lores (1998) have identified that the contractors play an important role in preparing schedule and budget, selecting major materials, construction methods, suggesting structural systems, if they are involved at the conceptual design phase.

Mitropoulos and Tatum (2000) showed concern about fragmentation of goals as one of the major issues that influenced the construction industry in recent days, which was a result of specialization of expertise. As a result, the successful and timely completion of project may suffer. In this situation, the main objectives were to develop integration framework. Nine managers were interviewed to derive at managerial techniques employed. The following benefits were identified:

- Improved project cost-effectiveness and schedule
- Increased safety
- Prevention of claims
- Improved logistics management and cash flows

It was also observed that integration is important at design phase for two important reasons: (1) to prevent problems in subsequent processes and (2) to select the alternatives that may optimize the project performance. It is important that contractors and vendors participate as "equal-partners" in design and joint decisions are done. Mitropoulous and Tatum (2000) have mentioned that as per some researchers, the process of integration requires exchange of information and knowledge between the independent subsystems. They also added that integration requires joint decision-making. The research concluded that owner has to take some important decisions regarding integration process such as selection of contractor may not be done at the lowest bid but rather focusing on his integration skills. Owner can train personnel for integration. Besides this, special incentives may be offered to parties actively participating for the project success. The benefits of integration in private sector are that in design stage it leads to the most effective solutions for cost saving and winning the contracts. The performance of such projects has an impact on further relationship of the contractor with the corporate client. In public sector, the contractor's previous performance and reputation is important in terms of his aggressiveness and confidence to bid, although the lowest bid is an important criterion.

Early involvement of contractor in design allows the contribution of construction knowledge and experience to design. Direct involvement of contractor gains better cooperation between the contractor and other participants throughout the design and construction process (Jergeas and Put, 2001). In another paper, Gil et al. (2004) mentioned the input of a contractor at early stage into four areas such as (1) ability to develop creative solutions, (2) knowledge of construction space needs, (3) knowledge of fabrication and construction capabilities, and (4) knowledge of supplier-led time and reliability. Othman (2011) has recommended for design firms to integrate construction knowledge and the contractor's experience in design process as approach to reduce construction waste and improve building performance.

2.3.2 Coordination

It is the backbone of efficient project management system. It has been observed that lack of coordination among team members can lead to unexpected delays in the project thus resulting in loss of time and money. Higgin and Jessop (1965) have studied the building construction industry and identified three main functions of the building process: the design, the construction, and the coordination. "Coordination is almost equivalent in meaning to control planning or management but is more descriptive of relating together of separate activities and their concerted direction towards a common purpose."

Crichton (1966) mentioned in Tavistock studies that the activity of coordination is carried out in an informal manner in the building industry. He further adds that coordination is not generally spoken of, on record. It does not appear in the handbooks or formal reports. O'Connor et al. (1987) suggested that interorganizational communication should be encouraged and planned, for particularly between designers and contractors. While defining constructability and total quality management, Russell et al. (1994) analyzed that both factors stress commitment from all personnel, i.e., from executive level to the level of the construction craftsmen at site. This process requires teamwork as an important tool. Coordination has also been defined as effective harmonization of planned efforts for accomplishing goals. It is the most important and sensitive issue of management. Coordination acts as a bridge in and fills up the voids created in various departments by changing situations in system, procedures, and policies (Chitkara, 1998).

Saram and Ahmad (2001) performed a research at identifying various activities that are performed to achieve coordination, which among those are most important and which among those are most time-consuming coordination activities. They identified 64 coordination activities and based on 33 responses received from practitioners in Hong Kong construction industry concluded the results. The six most important coordination activities have been identified as follows:

- Identifying strategic activities and potential delays

- Ensuring the timeliness of all work carried out

- Maintaining records of all drawings

- Information directives, verbal instructions, and documents received from the Consultant and Client

- Maintaining proper relationship with Client, Consultant, and Contractor

- Liaison with the Client and the Consultant

The activities that consume most of the time are identified as follows:

- Conducting regular meetings and project reviews

- Gathering information on requirements of all parties and consolidating for use in planning, resolving differences, etc.

The study also identified some important facts, like it is important to identify the activities which have greater impact than the other activities.

A paper by Carr et al. (2002) analyzed the importance of coordination during design phase of the project and highlighted that the interpersonnel interaction is important. This helps in integration of various components of the design. They further added that various professionals must interact with each another to bring together various components of the project in a coordinated fashion. Shen et al. (2004) stated that the multi-tier subcontracting system makes project communication and coordination difficult.

Jenitta and Tapadia (2004) have quoted that number of communication problems in the construction industry occur because of low coordination low efficiency, poor quality, and adverse attitudes. They further explored that Design and Build projects lead to better communication in the project team because all the team members work under single entity. All the parties are working for the same interest hence the communication is better. The working environment is productive and collaborative because the designers and contractors work simultaneously for single goal to provide the best solution to the client.

2.3.3 Bidding Process

The type of bid plays an important role in the successful accomplishment of the project. Many research projects have been carried out studies on different bidding processes.

Tatum (1990) identified the need for the early involvement of contractor in design. The Chartered Institute of Building has given the definition of Design and Build method that explains that the client directly deals with the contractor for completion of the building in Design and Build projects. It is further explained that the contractor is responsible for the coordination between design and construction processes. The consultants are free in such cases and the client may appoint either in house staff or a separate consultant to check that the contractor is providing value for money and that content and quality are satisfied (CIOB, 1988).

Glavinich (1995) discussed one of the problems of Design–bid–Build contracting system. The builder accepts the contract without asking for any kind of corrections in design and bidding time is short and the builder has little time to review. The builder later requests for extra time or extra compensation which appears to be an easy remedy, but later this can result in serious impacts like delays of projects or affecting the financial feasibility of the project. Pocock et al. (1996) outlined research looking at project interaction. The author discussed that "Most engineers and architects could benefit from contractor's input, but contractors are not usually involved in a project until bidding. They work from completed drawings and specifications without having any input to their contents." The contractors also agree that the traditional Design–bid–Build approach encounters the following difficulties: unrealistic schedules, specification problems, problems with physical interference, tolerance problems, and weather-related problems that could be avoided during design phase (Farooqui and Ahmed, 2008).

According to Mitropoulous and Tatum (2000), Design–Build contracting is the best and an effective mechanism to facilitate integration of design and construction. Three main types of mechanism were identified to increase the project integration: (1) contractual, (2) organizational, and (3) technological. Design–Build contracts have been suggested, as the entire responsibility of engineering, procurement, and construction process is under one organization. It is also appreciated because the contractors get an opportunity to participate in the design process right from the beginning of the project. The contractors give importance to corporate relationship and maintain long-term relationship with the designers. This helps them understand the needs of the client and win the contract, even if the bid is not lowest. The construction firms, which do not have in-house design cells, insist on maintaining relationships with the designers. Such relations help them gain projects through joint proposals also, at times.

In the traditional contracting practice, the contractor is selected through competitive bidding when the design has been completed by designer based on the knowledge that he has regarding aesthetics functionality, budget, and engineering consideration. In such cases, the contractor has little input to design. The construction knowledge and experience are important input for design, but their impact is limited in such cases (Arditi et al., 2002). Gil et al. (2004) emphasized that early involvement of the speciality contractors in the design process can be achieved by Design–Build contracting system.

Jenitta and Tapadia (2004) have explained the philosophy of Design–Build procurement method as "single point source." The Design–Build methodology provides best combination of design, construction, buildability, and economy. The Design–Build method has better scope of achieving synergy between the two phases of design and construction as compared to Design-bid–Build because in the previous case a single body is responsible for all the major decisions and activities with fewer conflicts. The advantages of Design–Build can be listed as:

- Shorter project execution time
- Single point responsibility
- Very less claims and disputes
- Greater privacy certainty
- Economy of project
- Better communication in the project team
- Collaborative work environment

The authors further added that Design–Build structures could be Designer-led Design–Build, Contractor-led Design–Build, and Novated Design–Build. In the third category, the client hires the designer and gets the design prepared. The contractors bid on this design and the successful Contractor enters into contract with the designer and develops the design details and executes the project.

Kansara et al. (2007) has shown in their research that vendors are selected by companies based on parameters that vary with the projects. Some of these can be listed as lead time, quality, response, and expenditure with the vendor.

2.3.4 Construction-Driven Schedule

The schedules are prepared in the construction projects to keep a check on various design and construction activities. Such schedules are expected to set effective guidelines for the timely completion of the project.

O' Connor et al. (1987) has discussed that constructability of a project is increased when the design and procurement schedules are construction driven. The construction schedules should be prepared even before the design and procurement schedules are finalized. This leads to reduced project duration, fewer delays in field, effective prioritization of various activities, effective work package, and goals of project are well known to the project personnel. Another paper by Glavinich (1995) explains that as the design process progresses the schedule must be updated on a regular basis. A Barr (Gantt) chart schedule should be prepared that identifies important activities. As the design progresses the schedule should evolve from initial bar chart to an informative network type chart schedule that shows activities and durations and their interrelationships. The design process is the time having much potential to correct the scheduling problems. The construction-driven schedules shall take into consideration the methods and techniques also that shall be adopted in the project execution. This decision at earlier stage saves time at the execution stage of the project, as the joint decisions are taken by the team members before the situation actually arrived on site.

2.3.5 Simplification of Design

Simplification of design is important, but this simplicity should not hinder the creativity of the designer. O'Connor et al. (1987) analyzed in their research that constructability is increased when designs have considered efficient construction, i.e., designs are configured to enable efficient construction. Some principles that can be adopted for simplifying designs are listed as follows:

- Use of minimum number of components, elements, or parts for assembly.

- Use of readily available materials in common sizes and configurations.

- Use of simple, easy-to-execute connections with minimum requirement of highly skilled labor, and special environment controls.

- Use of design which minimize construction task interdependencies.

Khan (2002) has discussed that some researchers have also identified the "Ten Commandments" for design which can help in increasing Constructability. Some of the important issues focus on keeping the design straight and simple in form, keeping the supports simple, standardization, keeping design site suitable and schedule sacred, etc.

The simplified design can increase the constructability of the building project. It is suggested that the design should be reviewed by qualified construction personnel.

2.3.6 Standardization of Elements

O'Connor et al. (1987) have discussed the importance of standardization in their paper and explained that Constructability is enhanced when the design elements are standardized, and repetition is followed. This also leads to savings because variations are minimized. Various areas where standardization can be applied are building systems, materials types, construction details, dimensions, and elevations. The extent to which standardization may be applied depends on the economic analysis also. The reduction in variety can lead to many benefits such as discounts on more of same material, simplified procurement, and materials management.

Another paper by Fischer and Tatum (1997) identifies some of the preliminary design variables which are important for constructability such as dimension of elements, distances between elements, their repetition, and modularity of layout. It is also suggested that the constructability can be improved at preliminary design stage in three types of design decisions: the horizontal layouts, vertical layouts, and the dimensioning of structural elements. Kansara et al. (2007) mentioned in their paper that "when a company sets up its own standards for the codification and own standardization of materials, it helps in the variety reduction as one can constantly monitor the amount of the materials used."

2.3.7 Prefabrication

Prefabrication is another method of achieving standardization. The different elements or sections of a building are cast in factory/off-site and they transported and assembled on-site. While discussing constructability, O'Connor et al. (1987) identified that ease of construction enhances if preassembly work is thought of in advance and preassembly/module designs are incorporated in advance to facilitate the process of fabrication, transport, and installation. It should be taken care of at the conceptual planning stage. The items which can be prepared off-site should be analyzed at an early stage of design. This can lead to many benefits such as improved task productivity, parallel sequencing of activity, increased safety, improved quality control, and reduced need for scaffolding. O' Connor et al. (1987) also studied that preassembly can increase constructability in case of elevated works because the need for scaffolding is reduced/eliminated. This issue is also helpful in situations where site is congested, and quality sensitive work is to be produced. Adverse weather conditions also promote the need for modular construction practices. Thus, the adoption of such practices can increase the project efficiency by saving important resources such as time and labor.

2.3.8 Accessibility to Site

Accessibility to site is a very important parameter because the ease of construction largely depends on this factor. O' Connor et al. (1987) addressed that the constructability enhancement can be achieved when the design promotes accessibility of manpower, material, and equipment. As study of many researchers highlighted that accessibility becomes very important and crucial in cases where the sites are tight, or roads capacity is limited, in case of renovation projects, working on high elevations, sites with steep grade changes, sites with extreme weather conditions or environmental conditions (like vegetation) or sites where multiple contractors are working. It is important to plan accessibility to site in terms of project elements, well defined and specified access lanes, and clear spaces for placement of equipment. Proper communication is required with designers regarding transport, erection and sizes of equipment in terms of clearances, etc.

Accessibility is important with respect to all the measures on the site to make the construction process easier and workable. Preplanning and well-thought-out design can lead to hurdle free and smooth construction of project. The construction methods, techniques, and equipment are to be decided at initial stage of design and considered at site layout stage. Proper considerations need to be taken for the execution as well as the operation and maintenance of the building also, during its life cycle.

2.3.9 Adverse Weather Conditions

Adverse weather conditions refer to the unfavorable climatic conditions or may be unexpected and unpredictable weather such as storm, fog, and snow. According to O' Connor et al. (1987), constructability can be increased when design facilitates construction under adverse weather conditions, in case they exist. This is crucial in countries where climate is a challenge for construction activities smooth functioning. Both the designer and constructor must be sensitive toward planning in such regions. Proper investigation is required to be done by the designer in advance to find out ways in which exposure to temperature extremes and effects of rain can be minimized. One of the major concerns in such cases is the quality control. Some of the important measures that can be incorporated are allowance for large enclosed spaces that can be used as fabricating shops and equipment storage, early paving of site to eliminate muddy operations, specifications such as admixtures for overcoming the effects of extreme weather and maximizing off-site work.

2.3.10 Simplification of Technical Specifications

Technical specification is related to detailed description of technical requirement in terms of suitability for design development of an item. Simplification of technical specifications reduces problems like over budgeting, unavailability of resource persons, poor workmanship, and project delays.

O' Connor et al. (1987) mentions that input should be invited from the construction personnel in finalizing of preferred specifications and methods but that should not be constraining design configuration. In case the views of construction personnel vary, specifications should allow for cost effective alternatives. Glavinich (1995) mentioned that the specification of special or custom equipment or material should be avoided. Also, the specification of obsolete materials, equipment, and construction techniques should be avoided.

2.3.11 Encouragement to Innovations

Encouragement to innovations refers to promotion of new ideas and it can be the application of better solutions that meet the new requirement.

Cox (1985) defined innovation as an attempt by "right people" to the demands of their job. It is defined as "Innovation is a byproduct of people who are acting on their unique strengths and who are refining their gifts." Foster (1986) explains that people who work for innovation are driven by higher project objectives and have a balanced perspective on change. Such people have an aggressive "attacker" approach and they are working on improving the inadequacies of current technology.

O' Connor et al. (1987) identified that good management practices should include practices like challenging of past practices and rewarding innovative ideas. They also mentioned in their paper that good ideas should be developed, and success should be documented. Further O'Connor et al. (1988) added that there are certain common innovation practices that can enhance the constructability of construction projects. These have been listed under various heads such as sequencing of field tasks, materials, and equipment etc. Some of the ideas are as follows:

- Sequencing of equipment such as crane, scaffolding, hoisting equipment, especially if they are to be used by multiple subcontractors. This will help reduce confusion and congestion on-site.

- Lighting systems may be installed at an early stage to reduce the need for temporary lighting.

- Stairs and platforms may be erected at an early stage. That may also help speeding up of work.

- Methods like steam curing, ground freezing are some advances in temporary construction systems.

- Innovations in formwork, such as flying formwork, ship form system are easily erectable.

- Advances in labor hand tools can increase mobility, accessibility, safety, and reliability such as cordless power hand tools, and automatic nailing gun.

- Constructability is also tending to make processes more of machine driven than worker driven. The processes can be speeded up with fully automated concrete batch plants, remote-controlled welding systems, automated concrete floor finishers, spray robot for structural steel fireproofing, etc.

- Temporary innovative facilities such as enclosures of work space in adverse weather with easier erectable tent, site pavement with easily available local material such as shells, etc.

2.3.12 Past Lessons Learned Exercise and Reviews

Past lessons learned exercise and reviews refers to the analysis or assessment of something adopted/performed in past, to be used for beneficial purposes in future.

O' Connor et al. (1987) writes that if the specifications are reviewed in detail by the designer, the owner and construction personnel, the constructability of project enhances, and field operations become simplified. Later O'Connor and Davis (1988) added that future chances for increasing the constructability can be thought of by documenting the preferences and innovative ideas of the constructors. This will help and benefit the future projects. Poor documentation work cannot be retrieved on time when required and can hinder the constructability. Proper information management systems should be taken care of by the designers as well as the constructors.

Russell et al. (1994) writes that maintaining a lessons learned database allows communication of positive and negative activities and experiences from one project to the future project. Glavinich (1995) made a mention of the term Design Phase Constructability Review and discussed that the design reviews should be conducted by senior design and field personnel prior to the start of the work which helps in promotion of better relationship between office and field personnel. The benefits of such reviews are increased goodwill, greater design constructability, and continuous scrutiny of the firm's design policies and standards.

Another paper by Fischer and Tatum (1997) concluded that often the corporate lessons learned are overlooked. Generally, there are no formal systems of keeping the feedbacks. It is important, and a formal system is required to acquire construction knowledge and to channel this knowledge to designers so that it can prove to be beneficial for the designers and contractors. The knowledge is collected during and after the construction phase of the project and the information is used as ready reference for other projects in future, so that those hindrances and problems are avoided.

2.3.13 Availability of Resources

Availability of resources refers to the ease with which various resources such as building material, labor, and equipment can be approached, hired, and put to work for the execution of the project. O' Connor et al. (1987) writes that it is always advisable to avoid materials which are difficult to obtain. A paper by Glavinich (1995) discusses that the architects and engineers should consider the available local material, conditions as well as construction practices. The availability of labor, material, and equipment should also be considered in design, i.e., the type of labor skills and construction practices which are not locally available should be avoided, so that the project cost can be controlled, and delays avoided.

The suggestions and experience of team members can be fruitful in selecting the most appropriate resources, but it is possible only when the integration of team is done at the initial stage of the project. Besides this the reviews of past lessons learned also guide in selection of resources, depending on their availability. The project may suffer if high-tech specifications are specified but the workmanship available is poor. This may result in poor quality of construction and heavy maintenance requirement during the operation of the building.

2.3.14 Appraise Recycling

Recycling means to convert waste into reusable products. Hemalatha et al. (2008) discussed construction and demolition waste and highlighted the importance of recycling. The construction and demolition waste is 10%–20% of municipal waste. The construction and demolition is said to be produced whenever any construction or demolition activity takes place. Such wastes are heavy, bulky, and need huge amount of space for storage. The authors have made a mention in their paper that according to the Technology Information Forecasting and Assessment Council (TIFAC, 2000), New Delhi, 70% of the construction industry is not aware of the recycling techniques. The construction and demolition waste management has been categorized into four stages: (1) storage and segregation, (2) collection and transportation, (3) recycling and reuse, and (4) disposal. A thorough understanding and encouragement of these practices shall increase constructability of the project.

2.3.15 Waste Management

Waste management means to organize and regulate the waste. Haghi (2010) defined waste management as "the collection, transportation, processing, recycling or disposal, and monitoring of waste material." The term is generally employed when referring to materials produced by human activity. Waste management needs to be done to recover resources from it and to reduce its impact on the health and the environment or aesthetics.

Earlier Kansara et al. (2007) have stated that waste is something that is unwanted and may be produced on the construction site or may be on the closure of the project. According to the authors, to increase the profits, it is very important to reduce the wastage, which is the indirect expenditure. Management software can help in keeping a check on the amount of material used in the project but generally the companies are not employing these methods and checking the waste manually on-site, which leads to time wastage. Waste management has not gained importance in Indian construction industry. Waste needs to be cut down to save economy. Government should set norms and standards for allowable waste percent. Based on severity, certain causes of waste have been identified.

The highly severe causes are as follows:

- Improper planning
- Poor management
- Improper quality control
- Lack of individual responsibility
- Overall negligence

The moderately severe causes are as follows:

- Improper designs
- Improper specifications
- Improper labor and supervision to faulty systems

The low severity causes are as follows:

- Lack of technological know-how
- Unavailability of resources
- Unhygienic working environment
- Lack of standardization

Waste management includes the management of waste generated on-site during construction and during its function also. A good design considers these wastes as important resources, in advance. The waste generated on-site can be reduced, reused, or recycled.

2.3.16 Application of Advance Information Technology

Employment of advance information technology refers to the adoption of latest and modern computerized means of technology for the project. O'Connor et al. (1988) observed that computer-aided design (CAD) overlay techniques have proven useful for studying the accessibility problems during the project execution, in advance. In some complex cases, computerized simulation models have been prepared to plan work flow and logistics. A paper by Fischer and Tatum (1997) concludes that CAD and expert system technology can also help in corporate knowledge like lessons learned from the projects, so that it can be applied at the design stage automatically and in this process the constructability of the project will be increased by higher quality of product. Such data system with past lessons learned information incorporated in their program, will help the designers save time and energy and make the project cost effective. Kansara et al. (2007) found that companies in India are using most commonly "MS Project" to plan out the quantities of material to be used. "PRIMAVERA" is also used by some companies to cross-check the planning done by other means. The design and construction activities can be regulated more effectively by employing the advance information technologies.

REFERENCES

Arditi, D., Elhassan, A., & Toklu, Y. C. (2002). Constructability analysis in the design firm. *Journal of Construction Engineering and Management, 128*(2), 117–126. DOI:10.1061/ (ASCE)0733-9364(2002)128:2(117).

Business Roundtable. (1982). Construction industry cost effectiveness project report. NY: Business Roundtable.

Carr, P. G., de la Garza, J. M., & Vorster, M. C. (2002). Relationship between personality traits and performance for engineering and architectural professionals providing design services. *Journal of Management in Engineering, 18*(4), 158–166.

Chitkara, K. K. (1998). *Construction Project Management: Planning Scheduling and Controlling.* New Delhi: McGraw-Hill.

CIOB. (1988). Design and build: Code of estimating practice supplement No.2: The Chartered Institute of Building (CIOB).

Cox, A. (1985). *The Making of the Achiever: How to Win Distinction in Your Company.* New York: Dodd Mead & Company.

Crichton, C. A. (1966). *Interdependence and Uncertainty: A Study of the Building Industry.* London: Tavistock Publications.

Farooqui, R. U., & Ahmed, S. M. (2008). Assessment of constructability practices among general contractors in Pakistan's construction industry. *CIB W107 Construction in Developing Countries International Symposium on Construction in Developing Countries: Procurement, Ethics and Technology.* Trinidad & Tobago, West Indies.

Fischer, M., & Tatum, C. B. (1997). Characteristics of design-relevant constructability knowledge. *Journal of Construction Engineering and Management, 123*(3), 253–260. DOI:10.1061/ (ASCE)0733-9364(1997)123:3(253).

Foster, R. N. (1986). *Innovation: The Attacker's Advantage.* New York: Summit Books.

Gil, N., Tommelein, I. D., Kirkendall, R. L., & Ballard, G. (2004). Theoretical comparison of alternative delivery systems for projects in unpredictable environments. *Journal of Construction Management and Economics, 22*(5), 495–508. DOI:10.1080/01446 190310001649100.

Glavinich, T. E. (1995). Improving constructability during design phase. *Journal of Architectural Engineering, 1*(2), 73–76. DOI:10.1061/ (ASCE)1076-0431(1995)1:2(73).

Haghi, A. K. (Ed.). (2010). *Waste Management: Research Advances to Convert Waste to Wealth (Waste and Waste Management Series).* Ottawa, ON: Nova Science.

Hemalatha, B. R., Prasad, N., & Subramanya, B. V. V. (2008). Construction and demolition waste recycling for sustainable growth and development. *Journal of Environmental Research and Development, 2*(4), 759–765.

Higgin, G., & Jessop, N. (1965). *Communications in the Building Industry: The Report of a Pilot Study,* London: Tavistock.

Itami, H., & Roehl, T. W. (1987). *Invisible Assets: Mobilizing Invisible Assets* (pp. 12–31). Cambridge: Harvard University Press.

Jenitta, S., & Tapadia, S. (2004). Design and Build (D&B): An integrated approach to construction projects. *NICMAR Journal of Construction Management, XIX*(II), 18–35.

Jergeas, G., & Put, J. V. (2001). Benefits of constructability on construction projects. *Journal of Construction Engineering and Management, 127*(4), 281–290. DOI:10.1061/(ASCE)0733-9364(2001)127:4(281).

Kansara, C., Reddi, S. A., & Shah, R. (2007). Identification of material wastage in residential buildings. *NICMAR Journal of Construction Management, XXII*(1), 13–34.

Khan, S. (2002). Designing for constructability and minimum mainte-nance. (Master's thesis). IIT Delhi, New Delhi, India.

Kichuk, S. L., & Wiesner, W. H. (1997). The big five personality factors and team performance: Implications for selecting successful product design teams. *Journal of Engineering Technology and Management,* *14*(3), 195–221. DOI:10.1016/S0923–4748(97)00010–6.

Mitropoulos, P., & Tatum, C. B. (2000). Management driven integration. *Journal of Management in Engineering, 16*(1), 48–58. DOI:10.1061/ (ASCE)0742-597X(2000)16%3A1(48).

Nam, C.H., & Tatum, C. B. (1992). Non-contractual methods of integra-tion on construction projects. *Journal of Construction Engineering and Management, ASCE, 118*(2), 385–398. DOI:10.1061/ (ASCE)0733-9364(1992)118:2(385).

Nima, M. A., Abdul-Kadir, M. R., & Jaafar, M. S. (2001). Evaluation of the role of the contractor's personnel in enhancing project con-structability. *Structural Survey, 19*(4), 193–200.

O'Connor, J. T., & Davis, V. S. (1988). Constructability improvement during field operations. *Journal of Construction Engineering and Management, ASCE, 114*(4), 548–564. DOI:10.1061/ (ASCE)0733-9364(1988)114:4(548).

O'Connor, J. T., & Miller, S. J. (1994). Barriers to constructability imple-mentation. *Journal of Performance of Constructed Facilities, ASCE, 8*(2), 110–128. DOI:10.1061/(ASCE)0887-3828(1994)8:2(110).

O'Connor, J. T., Rusch, S. E., & Schulz, M. J. (1987). Constructability con-cepts for engineering and procurement. *Journal of Construction Engineering and Management, ASCE, 113*(2), 235–248. DOI:10.1061/(ASCE)0733-9364(1987)113:2(235).

Othman, A. A. E. (2011). Improving building performance through integrating constructability in the design process. *Organization, Technology and Management in construction: An international Journal, 3*(2), 333–347. DOI:10.5592/otmcj.2011.2.6, Research paper.

Pocock, J. B., Hyun, C. T., Liu, L. Y., & Kim, M. K. (1996). Relationship between project interaction and performance indicators. *Journal of Construction Engineering and Management, ASCE, 122*(2), 165–176. DOI:10.1061/(ASCE) 0733-9364(1996)122:2(165).

Pulaski, M. H., Horman, M. J., (2005). Organizing constructabil-ity knowledge for design. *Journal of Construction Engineering and Management, ASCE, 131*(8), 911–919. DOI:10.1061/ (ASCE)0733-9364(2005)131:8(911).

Pulaski, M. H., Horman, M. J., & Riley, D. (2006). Constructability practices to manage sustainable building knowledge. *Journal of Architectural Engineering, 12*(2), 83–92. DOI:10.1061/ (ASCE)1076-0431(2006)12:2(83).

Radtke, M. W. (1992). Model constructability implementation procedures. (Master's thesis). University of Wisconsin, Madison, WI.

Russell, J. S., Swiggum, K. E., Shapiro, J. M., & Alaydrus, A. F. (1994). Constructability related to TQM, value engineering, and cost/ benefit. *Journal of Performance of Constructed Facilities, ASCE, 8*(1), 31–45. DOI:10.1061/(ASCE)0887-3828(1994)8:1(31).

de Saram, D. D., & Ahmed, S. M. (2001). Construction coordination activities: What is important and what consumes time. *Journal of Management in Engineering, 17*(2), 202–212. DOI:10.1061/ (ASCE)0742-597X(2001)17:4(202).

Shen, L. Y., Linda, F. C. N., Lucille, W., & Paul, F. (2004). Implementing environmental management in construction projects. *NICMAR Journal of Construction Management, XIX*(II), 1–17.

Tatum, C. B. (1987). Improving constructability during conceptual planning. *Journal of Construction Engineering and Management, ASCE, 113*(2), 191–207. DOI:10.1061/(ASCE)0733-9364(1987)113:2(191).

Tatum, C. B. (1990). Integration: emerging management challenge. *Journal of Management in Engineering, ASCE, 6*(1), 47–58. DOI:10.1061/(ASCE)9742-597X(1990)6:1(47).

TIFAC. (2000). Utilization of waste from construction industry, Department of Science and Technology. New Delhi, India. http:// tifac.org.in/index.php/8-publication/184-utilisation-of-waste-from-construction-industry.

Uhlik, F. T., & Lores, G. V. (1998). Assessment of constructability practices among general contractors. *Journal of Architectural Engineering, 4*(3), 113–123. DOI:10.1061/(ASCE)1076-0431(1998)4:3(113).

Wong, F. W. H., Saram, D. D. D., Lam, P. T. I., & Chan, D. W. M. (2006). A compendium of buildability issues from the viewpoints of construction practitioners. *Architectural Science Review, 49*(1), 81–90.

CHAPTER 3

Benefits, Barriers, and Awareness about Constructability

3.1 INTRODUCTION

The benefits of introducing constructability can be tangible and intangible. This section presents the benefits and barriers to constructability. Research conducted worldwide which focus on the awareness level of project participants and showcase the scenario in the construction industry are also presented here.

3.2 THE BENEFITS OF CONSTRUCTABILITY

Long time back, the construction projects were single handed by the master builder, who used to take care of the design as well as the construction activities for a project. There was a huge amount of integration in this process as the design and construction considerations were very well taken care of. The early decisions regarding construction materials and methods could improve design and increase the buildability of the project. With the increase in the specialization in the construction industry, the design and construction activities got separated considerably. With lesser concern and knowledge about each other's areas of specialization, the buildability got affected and the need to reassure and integrate the two processes of design and construction brought into picture the concept of constructability. Constructability is a value management tool developed as an attempt to bring closer the design and construction activities to the level of integration, once achieved by master builder (Russell et al., 1994).

Russell et al. (1994) have also discussed in their paper the qualitative and quantitative benefits of constructability. The quantitative benefits may be stated as reduced engineering cost, reduced schedule duration, and reduced construction cost in terms of labor, material, and equipment. The qualitative benefits may be listed as improved site accessibility, improved safety, reduced rework, increased communication, reduced maintenance cost, increased focus on common goal, increased construction flexibility, etc. There are many significant benefits of incorporating constructability program; for the constructors are paid off with more and steady construction (Gil, 2001) and for the designers in terms of better relationship with owner and contractor, lesser lawsuits, and good reputation (Arditi et al., 2002).

Further Arditi et al. (2002) identified and ranked the benefits of constructability in design firms as better relationship with clients and constructors, being involved in fewer lawsuits, building good reputation, professional satisfaction, and efficient design.

A survey was performed which highlights various benefits of constructability in various proportions, which can be listed as follows: minimizes contract change orders and disputes (89%), reduces project cost (82%), enhances project quality (81%), reduces project duration (70%), increases owner satisfaction (60%), enhances partnering and trust among project team (58%), and others (7%). Other benefits enlisted by the practitioners of the constructability was "safety." The respondents also cited problems that could be prevented by improved constructability. Five of them have been analyzed as change orders (23%), delays (20%), cost overruns (16%), conflicts and poor communication (15%), and requests for information or design errors (14%) (Pocock et al., 2006).

In another paper, Motsa et al. (2008) discussed that implementation of constructability leads to enormous benefits. The major benefits are in the areas achievement of better design, improved site management, and enhanced quality of the project.

3.3 THE BARRIERS TO CONSTRUCTABILITY

Barrier is an obstacle to proper communication. The researchers have enquired into construction industry and performed appraisal of various barriers to successful implementation of constructability. Construction Industry Institute (CII, 1987) has classified barriers to constructability into various categories such as general barrier, owner barrier, designer barrier, and contractor barrier.

General barriers are identified as follows: complacency with status quo "This is just another programme," "Right people" are not available, discontinuity of key project team personnel, no documentation of lessons learned, and failure to search out problems and opportunities.

Owner barriers are identified as follows: lack of awareness of benefits, concepts, etc.; perception that constructability delays project schedule; reluctance to invest additional money and/or effort in early project stages; lack of genuine commitment; distinctly separate design management and construction management operations; lack of construction experience; lack of team building or partnering; disregard of constructability in selecting constructors and consultants; contracting difficulties in defining constructability scope; misdirected design objectives and performance measures; lack of financial incentive for designer; gold-plated standard specifications; limitations of lump-sum competitive contracting; and unreceptive to contractor innovation.

Designer barriers are identified as follows: perception that they have considered it, lack of awareness of benefits, concepts, etc.;

lack of construction experience/qualified personnel setting company goals over project goals; lack of awareness of construction technologies; lack of mutual respect between designers and constructors; perception of increased designer liability; and construction input is requested too late to be of value.

Contractor barriers can be listed as follows: reluctance of field personnel to offer pre-construction advice, poor timeliness of input, poor communication skills, and lack of involvement in tool and equipment development.

O'Connor and Miller (1994) added some more barriers to constructability: (1) *organized barriers* such as preassembly limitations and other work restrictions; (2) *vendor barriers* such as fragmentation and difficult communication interfaces and restrictions on proprietary designs; (3) *code authority barriers* such as rigid, outdated codes and design standards and non-rigorous approach to establishment of tolerances; and (4) *research barriers* such as difficulty in proving the economics of constructability.

Barrier to constructability is that impediment that stops effective implementation of constructability program. O' Connor and Miller (1994) assessed the barriers through in-depth interviews of representatives from 62 companies which claimed to have been using constructability programs. They identified the most problematic barriers to effective constructability improvement. These are as follows: (1) complacency with status quo, (2) reluctance to invest additional money and effort in early project stages, (3) limitations of lump-sum competitive contracting, (4) lack of construction experience in design organizations, (5) designer's perception that "we do it," (6) lack of mutual respect between designers and constructors, (7) construction input is requested too late to be of value, and (8) beliefs that there are no proven benefits of constructability.

On close examination of the research work conducted, it can be summarized that each team member has potential to contribute to the success and constructability of the project. Each participant has different roles and responsibilities to carry, during different phases of the project. The key to success is the timely action. The study of research suggests that the team members should possess the quality of reception with positive attitude to increase constructability in a project.

Certain barriers that contribute to gaps in achieving the benefits from constructability are lack of trust and credibility among team members and lack of desire by owner to commit resources to implement constructability. The stakeholders cannot foresee the innumerable benefits of implementing constructability and how their decisions at early stage of planning will affect field operations. They perceive the contractors as "doers" only and not as contributors to design and planning. Simultaneously, the contractors are also uncomfortable in office environment and would be hesitant in sharing their views unless invited. Barriers that restrict the efficiency in constructability efforts are congestion around construction site and rigid specifications that sometimes limit the flexibility of design. This happens because the alternatives and availability of resources may not have been thought off by the designer due to lack of his practical field experience. Major barriers that restrict the benefits from the application of innovative methods and advanced technology have been identified in a survey and fall into three areas: (1) risk aversion and lack of trust by owners, lack of knowledge of the latest construction methods and techniques, and the paradigms of "we have never done that before" and "this is what we did on the last job and it worked then, so why do something different now"; (2) the real or perceived high cost of advanced computer technologies, especially in field locations requiring sophisticated telecommunications links; and (3) the time required to adequately train staff in the use of computer systems that seem to change very frequently and the lack of user-friendliness (Jergeas and Put, 2001).

The most commonly cited obstacle to implementing constructability, identified during a survey, are as follows: lack of open communication between designers and constructors (64%), inadequate construction experience (45%), difficulty coordinating disciplines (44%), lack of resources (42%), project delivery methods (27%), contract type (25%) not part of current process (21%), too costly (13%), inconsistent terminology (13%), lengthens project (6%), and others (11%). The survey also highlights some other issues as obstacles: the codes do not require constructability, the curriculum does not specifically focus on constructability education, and designers lack experience and are defensive. However, approximately half the designers have mentioned that "Lack of adequate construction experience" is a major obstacle to implementing constructability, which makes it very clear that the awareness level of designers, is increasing, and they are trying to incorporate the constructability principles in the project. The builder's response agrees to the responses mentioned above, as they cited lack of open communication between designers and builders was the largest obstacle to constructability (22%) (Pocock et al., 2006). It is very likely that with the increasing complexity in the building projects an early involvement of constructor in the project can save resources.

3.4 GLOBAL AWARENESS LEVEL

Many researchers have tried to explore the level of awareness about the constructability, among the key role players of the construction industry. An examination of such studies shall help in identifying the current scenario in construction industry and help in recognizing gaps in the present literature.

Uhlik and Lores (1998) indicated that 90% of general constructors, whom they surveyed, did not have formal constructability programs. They did not either act toward its implementation.

Cox and Thompson (1998) surveyed 332 construction projects in the United Kingdom, and found that Design–Build (DB) contracting is 12% faster than the traditional designing and procurement methods. They are 13% cheaper and 50% more likely to finish on time.

Arditi et al. (2002) in the United States found that most design firms perceive the concept of constructability to 95.7%. Almost 50.7% of respondents have formal corporate philosophy about constructability in their organization. The author also indicated that 87% of surveyed design firms used constructability reviews during developed design stage. They stated that 95% of the respondents believed that construction engineers should be involved in the design phase, in addition to other professionals, who are already participating at this stage. Of these, 57% respondents believed that they should be involved, regardless of project conditions, whereas 38% indicated that the involvement should depend on size, complexity, and type.

Pocock et al. (2006) have shown that constructability has gained importance and it is increasingly being adopted and applied in early project stages and have discussed an online survey which was conducted by U.S. Army Corps of Engineers Construction Engineering Research Laboratory (2003) among approximately 100 respondents including owners, architects, engineers, consultants, constructors, and construction managers from the United States. The respondents said that constructability begins in pre-project planning (18%), in conceptual design (41%), and during detailed design (24%). A total of 35% respondents use constructability review and 29% prefer checklist to avoid common construction errors as mechanisms to address constructability on projects. Construction expert was a member of design team in 33% of responses and reviewed the design in 57% of responses. Fifty-nine percent of designers reported that constructability efforts were initiated at conceptual design stage whereas only 28% saying it begins during detailed design stage. The designers used a combination of these methods and the frequency is as follows: design review by a construction expert (59%), peer reviews (53%), a constructability review activity on your project schedule (38%), a construction expert on design team (34%), and implementing a database or checklist to avoid common construction errors (34%).

Motsa et al. (2008) in South Africa identified that 84% design firms are aware and perceive the concept of constructability. Seventy-six percent of the design firms indicated that they required constructors experience in their design because they have better knowledge about material availability and appropriate technology that affects design and construction.

Othman (2011) has mentioned a survey study in his paper that was conducted in South Africa, and it was found that 84% design firms were aware of constructability concept. Seventy-six percent of the firms indicated that they require constructors experience in their design because they had better knowledge about material availability and application technology that affects design and cost. All the respondents agreed that structural engineers were most commonly involved professionals. About 44.7% respondents stated that specialist subconstructors were least involved.

Kamari and Pimplikar (2012) conducted a survey of four construction companies in India and identified that most of the time the problem occurred with drawings because a thorough review was missing. The best type of contract was Build-Operate-Transfer (BOT) and Design Build (DB) as they had less number of constructability issues. It was found that 25% of the respondents performed constructability analysis throughout the entire design process (from conceptual to the finishing of design). It was also observed that 51% firms start performing reviews as early as conceptual planning stage. The most significant factor (87%) that affects constructability was project complexity. The second highest factor (75%) was design practices and philosophy, i.e., designers approach to problem which includes their attention to construction details, site experience, etc. The three important factors that were found to cause constructability problems were faulty ambiguous or defective working drawings and adversarial relationships. The respondents listed the most magnificent benefit of constructability reviews to design firms as: better relationship with contractor and client (83%) and reduction in lawsuits and number of claims (72%). They also performed a survey related to architectural designs and constructability issues directly. There were many architectural aspects which were to be rated as constructability issues. The most significant factor was architectural drawing (95%), compatibility between interior and exterior designs (75%), architectural new styles and shortage of enough knowledge (75%), shape of structure (65%), and procedure of developing architectural designs (65%), architectural design and acoustic solutions (65%), and materials chosen by architects (60%).

Pulaski et al. (2003) conducted a research and illustrated that merging of sustainability and constructability efforts can lead to synergies that produce significant improvements to project performance and quality. Another paper by Pulaski et al. (2006) studied constructability practices to manage sustainable building knowledge. They identified 35 ideas during study, but 14 were eliminated. Finally, 21 principles were categorized into three different groups. These principles are for achieving synergy between sustainability and constructability. The study focused on the renovation project of Pentagon and the relationship of the two issues was theoretically determined. These principles are listed in the following three categories:

Sustainability and constructability principles for design

1. Simplify and standardize construction details.

2. Standardize repeatable components.

3. Ensure proper sizing and specification of equipment, products, and materials.

4. Consider alternative water conservation and site drainage solution.

5. Simplify and separate building systems and components to facilitate maintenance and future renovations.

6. Consider construction worker safety and efficiency during design.

7. Use structural elements as finished materials.

Sustainability and constructability principles for design and construction

8. Reduce area disturbed during construction.

9. Optimize dimensions to utilize entire product/material.

10. Continuously search for alternative environmentally safe products/finishes.

11. Reuse construction materials, existing finish materials and products.

12. Use local materials and construction methods.

13. Use methods and materials that allow for ease of reconfiguration, renovation, and deconstruction.

14. Select fittings, fasteners, adhesives, and sealants that allow for quicker disassembly and facilitate the removal of reusable materials.

15. Minimize the use of all building components and materials.

16. Minimize piping and ductwork bends.

17. Prefabricate building components and/or modularize construction.

Sustainability and constructability principles for construction

18. Sequence construction activities to reduce unnecessary design requirements and minimize contaminant sinks.

19. Protect indoor air quality during construction to alleviate problems during construction and expedite building turnover.

20. Salvage and donate unwanted materials.

21. Reduce packaging waste.

Another study was conducted in Pakistan to assess the constructability practices among general contractors. Hundred questionnaires were sent, and the response rate was 64%. The firms were medium- to large-sized organizations and survey was completed by top management involved in quality management program having over 10 years of experience in project management and coordination activities in their organization. Almost one-third of the contractors believe that constructability efforts should begin during construction phase although all of them agreed in principle, to the definition of constructability, as defined by CII. A total of 70% of the contractors were in favor of applying the constructability to the projects of all nature and sizes. The potential benefits of applying constructability have been identified as: reduce project duration (90%), enhance project quality (73%), enhance project safety (57%), reduce project cost (53%), minimize contract change orders and disputes (50%), enhance partnering and trust among project team (50%), increase stakeholder satisfaction (47%). The participation of contractor's during pre-construction activities for constructability implementation is 25%, which is due to traditional Design–bid–Build approach. However, the contractor's involvement during filed operation activities for constructability implementation is 62%. Certain other issues focusing on implementation of constructability in the organization were also surveyed. The survey results show that the constructability is seldom required by owner (24%) and architects (18%). The top management supports constructability (67%), but only less than one-third of organizations have an organized constructability program (Farooqui and Ahmed, 2008).

It is important and essential that there is teamwork and it is focused on all the aspects of building, right from the design stage to the construction and operation stage of the building project.

In the conventional approach, the design practices lack integration because of which it is difficult to reach appropriate solutions.

Integrated design in not necessarily "high-tech" or specialized technical design. The focus is on long-term functioning and health of an entire building system, not just specific elements. Integrated design is not sort of traditional "handoff" or sequence of activities proceeding linearly from owner to architect to engineer to general contractor to subcontractor to occupant; instead, there are built in feedback loops as each step of design is evaluated against project goals. Elements of the integrated design process are as follows:

- Make commitment to integrated design and hire design team members, who want to participate in a new way of doing things.
- Set "stretch" goals for the entire team.
- Get the team to get to zero cost increase over a standard budget.
- Front load "the design process with environmental charrettes, studies and similar thinking time."
- Allow enough time for feedback and revisions before the final design concept.
- Everyone must participate.

The role of contractor in the integrated design process is very important and no discussion would be complete without examining his views and experience. The constructors coordinate the work of dozens of trades and subconstructors and are directly responsible for spending more than 90% of the projects budget (Yudelson, 2009).

Another online survey was conducted, and data collected from 106 architects in India. Forty-one percent of them are practicing the profession and are non-academicians, 50% architects were academicians, and 9% architects were exclusively professionals, practicing project management. All 77% of architects were practicing the profession. The survey included 18 questions on various issues of constructability to analyze the present scenario of constructability practices in Indian construction industry. Some of the results are as follows: The construction personnel are considered as team member in conceptual design stage in 18.7% responses, at design development stage in 31.8% responses, and at field operations stage in 95.4% responses. The consultant is considered as team member in conceptual design stage in 62.3% responses only. Only 48.6% respondents agreed that concern was on increasing the site efficiency. Review meetings are regularly conducted in 41.7% responses at the conceptual planning stage, in 71.3% responses at the design development stage, and in 82.4% responses at the field operations stage. The concern at design development stage to reduce scaffolding during project execution was responded 35.5%. Other issues of field operation stage were responded as follows: waste management on site (44.4%), integration of contractor (95.4%), freedom to contractor (85.7%), and documentation work (58.9%). As the project progresses, the inclusion of project participants takes place and regular review meetings take place (Khan, 2016).

An overview of the awareness level of project participants highlights the following facts:

- The concept of constructability is not very popular, especially among the contractors.

- There is a preference for the DB type of project contracts, as the coordination and performance level is high in such projects.

- The importance of the concept of constructability is realized and practiced well in countries such as the United States, the United Kingdom, and South Africa.

- In case of the survey conducted in Indian construction industry, only 25% of the population of the sample size was practicing constructability.

- It has been accepted that the DB contracts involve less issues.

REFERENCES

Arditi, D., Elhassan, A., & Toklu, Y. C. (2002). Constructability analysis in the design firm. *Journal of Construction Engineering and Management, 128*(2), 117–126. DOI:10.1061/(ASCE)0733-9364(2002)128:2(117).

Construction Industry Institute. CII. (1987). Constructability concepts file. [Report]. Austin: University of Texas, Bureau of Engineering Research.

Cox, A., & Thompson, I. (1998). *Contracting for Business Success.* London: Thomas Telford.

Farooqui, R. U., & Ahmed, S. M. (2008). Assessment of constructability practices among general contractors in Pakistan's construction industry. *CIB W107 Construction in Developing Countries International Symposium on Construction in Developing Countries: Procurement, Ethics and Technology*, Trinidad & Tobago, West Indies.

Gil, N. (2001). Product-process development simulation to support specialty contractor involvement in early design. (Doctoral dissertation). University of California, Berkeley, CA.

Jergeas, G., & Put, J. V. (2001). Benefits of constructability on construction projects. *Journal of Construction Engineering and Management, 127*(4), 281–290. DOI:10.1061/(ASCE)0733-9364(2001)127:4(281).

Kamari, A. A., & Pimplikar, S. S. (2012). Architectural design and constructability issues. *AKGEC International Journal of Technology, 3*(1), 8–17. www.akgec.in/journals/Jan-June11/3-Ali%20.pdf.

Khan, S. (2016). A study of design and construction practices of sustainable architecture in India based on the concept of constructability. (Doctoral thesis). AMU, Aligarh, India.

Motsa, N., Oladapo, A. A. & Othman, A. A. E. (2008). The benefits of using constructability during the design process. *Proceedings of the 5th Post Graduate Conference on Construction Industry Development* (pp. 158–167). Bloemfontein, South Africa. www.researchgate.net/publication/237544357_the_benefits_of_using_constructablity_during_the_design_process.

O'Connor, J. T., & Miller, S. J. (1994). Barriers to constructability implementation. *Journal of Performance of Constructed Facilities, ASCE, 8*(2), 110–128. DOI:10.1061/(ASCE)0887-3828(1994)8:2(110).

Othman, A. A. E. (2011). Improving building performance through integrating constructability in the design process. *Organization, Technology and Management in Construction: An International Journal, 3*(2), 333–347. DOI:10.5592/otmcj.2011.2.6, Research paper.

Pocock, J. B., Kuennen, S. T., Gambatese, J., & Rauschkolb, J. (2006). Constructability state of practice report. *Journal of Construction Engineering and Management, ASCE, 132*(4), 373–383. DOI:10.1061/(ASCE)0733-9364(2006)132:4(373).

Pulaski, M. H., Pohlman, T., Horman, M. J., & Riley, D. (2003). Synergies between sustainable design and constructability at the Pentagon. *Proceedings of Construction Research Congress*, (pp. 1–8) ASCE, Hawaii.

Pulaski, M. H., Horman, M. J., & Riley, D. (2006). Constructability practices to manage sustainable building knowledge. *Journal of Architectural Engineering, 12*(2), 83–92. DOI:10.1061/(ASCE)1076-0431(2006)12:2(83).

Russell, J. S., Swiggum, K. E., Shapiro, J. M., & Alaydrus, A. F. (1994). Constructability related to TQM, value engineering, and cost/benefit. *Journal of Performance of Constructed Facilities, ASCE, 8*(1), 31–45. DOI:10.1061/(ASCE)0887-3828(1994)8:1(31).

Uhlik, F. T., & Lores, G. V. (1998). Assessment of constructability practices among general contractors. *Journal of Architectural Engineering, 4*(3), 113–123. DOI:10.1061/(ASCE)1076-0431(1998)4:3(113).

Yudelson, J. (2009). *Green Building Through Integrated Design*. New York: McGraw-Hill.

CHAPTER 4

Role of Constructability in the Life Cycle of Buildings

4.1 INTRODUCTION

The life cycle of a building needs special concern during the conceptual planning and design development stages because the decisions taken at this stage will have an impact on the building maintenance during post occupancy. The life cycle of a building is analyzed in this chapter in the following categories: (1) the conceptual planning stage, (2) the design development stage, (3) the field operations stage, and (4) the maintenance stage. The importance of constructability issues is discussed with respect to these stages of building's life span.

4.2 THE LIFE CYCLE OF BUILDINGS

The life cycle of buildings can be analyzed from the beginning of the project until the end of useful life of the structure. When the life cycle assessment of a building is performed, it involves whole life cycle of the building, which also includes the product stage of raw materials that would be employed in the building construction, and on the other end, it also talks about the benefits that could be extracted from the reuse and recycling of the materials.

However, our concern here is on the management issues related to constructability. Therefore, the thought process about the life cycle of a building shall begin with the idea of initiating the project and bringing it into existence. It is important for all team members to concentrate on all phases of the building cycle for best management and maximum energy saving potential. The designers should consider recycling, reusability, and deconstruction ease as an integral part of design during the initial planning stage of the project. The life cycle of a building also includes the usage period of a building, which means that maintenance is also an important consideration at the planning stage. The selection criteria for various components of structure should lead to ease of maintenance also. To summarize, the life cycle of a building can be broadly divided into four categories for study: (1) conceptual planning stage, (2) design development stage, (3) field operations stage, and (4) maintenance stage.

The following activities have been shortlisted under different heads, after the synthesis of literature.

4.2.1 The Conceptual Planning Stage

The conceptual planning stage is the first stage in the life cycle of a building. The team formulation should take place at the beginning of the project. The shortlisting of contractors can be suggested, or a panel of contractors can be of great help because their experience and input can lead to great savings in cost. There should be discussions on how to make the construction process easier amongst the team members and the valuable practical input of people with construction experience shall be of great help in this activity also. All the necessary surveys are to be conducted at this stage of planning. The construction methods are to be discussed and decided by the team members. This shall guide the site planning and solve the accessibility issues beforehand. Working out a construction schedule is important as a guideline for the further development and proceeding of the project. Laying out site efficiently is another important aspect which is well taken care of at this stage if all the team members share their input and experience. Discussion on recycling and simplification of technical specifications are other important discussions. Again, the input from construction personnel are of extreme importance because of their experience in the material availability, workmanship, etc. Review by all team members and implementation of past lessons learned is to be incorporated and adopted in a formal way to make the project a success.

4.2.2 The Design Development Stage

The design development stage is the second stage in the life cycle of a building. The development of design and procurement schedule is an important tool in managing the project. The final selection of the contractor and his team should be done so that whole-hearted involvement of the construction personnel can bring benefit to the project in the long run. The contractors once involved at this stage shall build up their team of subcontractors and vendors accordingly for smooth running of the project and start attending the review meetings. The application of advance technologies at the initial stage of a project minimizes project delays and cost overruns at a later stage. These should be utilized as an aid for improved management practice and control on project. The standardization of design elements shall help in faster construction practices and savings. It also helps in maintenance of the building during its occupancy. Small-scale physical models and 3D drawings shall be prepared as they enhance the capability about the visualization of forthcoming issues regarding accessibility on building site during construction and during building occupancy also. It is extremely important for team members to review the design at regular intervals. Such meetings are generally headed by the Project manager and are organized to discuss the progress and hindrances. The head of the meeting is responsible for noting of minutes and assigning duties and responsibilities and evaluating the scheduled progress. Such meetings are essential at all stages of work and help minimize issues of coordination among team members. Considerations are required for site drainage and water conservation during the entire life cycle of the project. Concerns to reduce scaffolding lead to better constructability because of faster and easier construction process. It also leads to savings in terms of cost and time. Thorough study and care is required while writing the specifications as everything that is constructed demands maintenance also. Due concern is recommended on this issue before planning and specification writing in context with availability of materials and workmanship, renovation work, and demolition of the building after its useful life is over. Environmentally safe materials and methods of construction should be adopted but unfortunately the construction personnel are not given enough freedom and opportunity to share their views and spend resources on these issues.

4.2.3 The Field Operations Stage

Field operations stage is the third stage in life cycle of a building. Usually, all the team members are involved in this stage of work and review meetings and inspections take place regularly. But it is essential to organize these meetings in a formal manner so that all the lessons learned are recorded and guarded for future adoption by some or all the team members as guidelines of dos and don'ts. Usage of temporary materials and systems on site can enhance the working efficiency and manageability. The contractors should be given freedom on site for valuable technical inputs for improving the construction process and innovations in equipment used. Their suggestions such as usage of precast systems can bring benefits in terms of saving time and cost during adverse weather conditions. Waste management should be an integral component of projects. Documentation work of the lessons learned during the project execution stage helps in recalling the errors to avoid repeating them in future. Such records should be discussed in regular inspection/meetings on/off-site by team members to avoid repetitive troubles.

4.2.4 The Maintenance Stage

The post-occupancy period of building demands maintenance during its life span for its safety and to maintain a healthy environment. To minimize maintenance problems during post-occupancy, the designer as well as the constructor need to be careful regarding various issues during the design and construction phase of the building. The choice of building material according to the site conditions and the available workmanship are crucial factors that can affect the quality of construction and later cause maintenance problems. Coordination between the designer and the constructor is the possible solution to this problem. The general discussions between team members regarding availability of building material and workmanship, regular review meetings, and visits to the site can help in reducing the maintenance problems at a later stage. It is also important to specify good details so that assumptions and ambiguities are reduced during field operations stage. Proper detailing and well-thought-out design is requisite for less maintenance. Thus, the concern for constructability issues can help in minimizing the maintenance problems of the buildings also.

CHAPTER 5

Case Studies in the Indian Context

5.1 INTRODUCTION

Case studies are an important tool and means of quantitative study. This section focuses on two studies based on observation, structured questionnaire survey, and scheduled interviews. The study is focused on Indian construction industry and purposive sampling is adopted to shortlist ten alternatives for study in Delhi and National Capital Region. The responsive group was a combination of architects, consultants, and project managers.

The purpose of study for the first study includes the following:

- Examination of the relationship between constructability and sustainability.

- Analysis of constructability practices followed.

- Identification of the challenging issues and project management systems adopted in the construction industry.

Another study was conducted with the help of structured questionnaire among 30 respondents who practice architecture.

A comprehensive list of 30 activities was prepared, organized in three different categories: conceptual planning stage, design development stage, and field operations stage. These activities are based on constructability practices. The respondents were requested to suggest the relationship among these activities, in terms of dependency or independency, with respect to other activities in the same category. The purpose of study was to identify the interrelationship among different design and construction (D&C) activities based on constructability. This information collected is further utilized to establish a sequential network of activities for all the above mentioned categories.

5.2 ANALYSIS OF STUDY 1

5.2.1 Task 1—Relationship between Constructability and Sustainability

This section focuses on the examination of relationship between constructability and sustainability. As the world is facing issues of sustainability, it becomes important to analyze constructability in the light of sustainable development. The research question is, "Whether there is a relation between constructability and Sustainability?" If yes, "to what extent." A positive relationship would help in enforcing the idea of constructability, its promotion, acceptance, and implementation in projects for achieving sustainable development.

5.2.1.1 Sustainability Study

Historical buildings made use of natural resources of energy and were comfortable for all seasons. The traditional methods of construction have always employed renewable resources of energy and local building materials. This environmental degradation increased because of greediness of mankind and exploitation of natural resources, hence the need to recall the concept of sustainability.

The sustainability has been an issue of great concern since the 1970s. An international think tank "The Club of Rome" was found in 1968, and in 1972 they published "The limits of Growth." The idea was that the economic development must be combined with environmental protection. The first UN Summit on man and environment took place in 1972. Gro Harlem Brundtland's report, "Our Common Future," was discussed in the 42nd UN Congress in 1987. In the 1990s, the United Nations Rio Earth Summit raised the issues regarding global warming, and the problems of ecosystem were also discussed here. The Inter-Governmental Panel on Climate Change brought to notice that during the 20th century, the Earth warmed up by between 0.3°C and 0.6°C, and sea levels rose to an average by 15–25 cm. The concept of sustainable development was introduced, and during the Rio Earth Summit in 1992, the heads of state committed that they shall explore "Development which fulfils current needs without compromising the capacity of future generations to fulfil theirs" (Muller, 2002).

A sustainable building is also defined as a building that (TERI, 2004)

- minimizes the use of natural resources during construction and operation
- uses efficient building material
- optimizes the use of on-site sources
- uses minimum energy for its working
- maximizes use of renewable sources of energy
- uses efficient waste and water management practices
- provides comfortable and hygienic indoor working conditions

According to the United Nations, the proportion of the world's population living in urban areas shall be greater than the population of those living in rural areas. It is expected that almost all the increase in the world's population during 2000–2030 will occur in urban areas, and about half of this will be absorbed by the urban areas of less developed regions. The evidences are saying that unless the emissions are reduced by at least 60% by 2050, the Earth's climate could move to an irreversible phase of global warming (TERI, 2004).

Sustainable development does not require a different method of construction. Any building that is sustainable does not essentially need to look different, beautiful, or possess a distinct identity. It is unique characteristic of the building, which can be realized by its working efficiency and the remarks of its end users. Such buildings will use optimum resources, be workable, consume less energy, use renewable sources of energy, and bring comfort to its users.

The architectural practices should have the ability to cope with the changes. The flexibility of design is important to stand with the changing demands of the society. It needs an assurance and adaptability. There should be sufficient place and access for every all. The idea of inclusion is important. The scarcity will bring social division and disturb the social sustainability. The feeling of ownership and belongingness is important. *When architecture is the outcome of all these ideas, it is sustainable architecture.*

5.2.1.2 Enlisting Sustainability Activities

After synthesis of the available literature, the important features of sustainable architecture were identified in five categories. Each category was further divided into five subcategories.

5.2.1.2.1 Efficient Site Planning

1. Preserve and plant vegetation on-site

2. Top soil preservation on-site

3. Preserve existing topography on-site

4. Terrace gardening

5. Microclimate control with water bodies

5.2.1.2.2 Water Conservation

1. Permeable ground surface

2. Reuse of waste water on-site by recycling

3. Low flow faucets/fixtures

4. Rain water conservation/reuse

5. Landscape using native species

5.2.1.2.3 Renewable & Waste Resource Management

1. Solar photovoltaic systems for lighting

2. Use of fly ash

3. Solar photovoltaic cells for water heating

4. Using precast systems

5. Waste management on-site

5.2.1.2.4 Building Design

1. Orientation of building

2. Courtyard planning

3. Daylight in interiors

4. Basement or underground structures

5. Shading devices/methods

5.2.1.2.5 Building Materials & Finishes

1. Building materials of low embodied energy

2. High-performance glass usage

3. Double glazing for insulation

4. Exterior finishes in white/reflective/brickwork

5. Low volatile organic compound paints in the interiors

5.2.1.3 Enlisting Constructability Activities

After synthesis of the available literature, the important activities of constructability are also identified in four categories, which are further divided into seven subcategories each:

- Database

- Category A—Conceptual planning stage

- Category B—Design development stage

- Category C—Field operations stage

5.2.1.3.1 Database   It is focused on the team work, integration, and coordination strategies.

1. Project team formed before the conceptual design

2. Architect was the team member

3. Project manager was the team member

4. Contractor was the team member

5. Consultant was the team member

6. The criterion for selection of team members was experience in similar projects

7. Contract was the Design–Build contract

5.2.1.3.2 Category A—Conceptual planning stage   It is focused on information collection and the involvement of construction personnel at the initial stage of project.

1. Conducting surveys

2. Discussion on construction methods

3. Design/project schedule

4. Ease of field operations

5. Discussion on recycling

6. Simplification of technical specifications

7. Review and implementation of past lessons learned

5.2.1.3.3 Category B—Design development stage   It is focused on design and planning strategies that ease the construction activities.

1. Use of advance information technology

2. Standardization of design elements

3. Review of design by other team members

4. Considerations for site drainage

5. Considerations for water conservation

6. Preference of methods for renovation and deconstruction

7. Consideration for environmentally safe and local construction methods and materials

5.2.1.3.4 Category C—Field operations stage   It is focused on the impact of involvement of team members at initial stage of design and freedom for innovation.

1. Field task sequencing (Critical Path Method/Project Evaluation and Review Technique)

2. Innovation in available equipment

3. Freedom to contractors for technical input

4. Considerations for adverse weather

5. Documentation work of the lessons learned

6. Waste management on-site

7. Regular inspection/meetings on/off-site by consultants

5.2.1.4 Analytical Study

Guttman scale is employed for scoring the sustainability features in five categories. Guttman's method of scaling is known as "scalogram analysis." The items for the scales are chosen such that they can represent a one-dimensional scale (Singh, 2008).

The identified ten case studies are analyzed on the above mentioned parameters and Decision matrix is prepared using Guttman's scale. The percentages calculated in the Decision Matrix for Sustainability and Constructability are used for the study of relationship between the two variables, as shown in Tables 5.1 and 5.2.

Analytical study was conducted to study relationship between the two variables: Sustainability (dependent) and Constructability (independent). Pearson's correlation coefficient and regression analysis were applied to the data and following was determined. The value of Pearson's correlation coefficient is 0.714. The p-value $= 0.02$ which is significant because $p < 0.05$. This implies that the result is significant with 95% level of confidence. There is a strong evidence for acceptance that *the application of Constructability principles shall increase the Sustainability of buildings.*

- The regression equation is given as follows:

$$Sustainability = 9.2 + 0.775 Constructability$$

This equation explains that one unit increase in the variable Constructability shall increase the variable Sustainability by 0.775. The regression analysis result is significant as the p-value $= 0.002 < 0.05$. There is a significant positive relationship between the two variables/factors: Sustainability and Constructability, as shown in Figure 5.1. Hence, it can be

TABLE 5.1   Decision Matrix for Sustainability Issues

S. No.	Criteria Case Study	Efficient Site Planning	Water Conservation	Renewable & Waste Resource Management	Building Design	Building Materials and Finishes	Total Score (Max. 25)	%
1	Alternative 1, Gurgaon	2	4	3	5	3	17	68
2	Alternative 2, Gurgaon	4	2	2	5	2	15	60
3	Alternative 3, Gurgaon	3	2	2	5	3	15	60
4	Alternative 4, New Delhi	2	2	2	4	2	12	48
5	Alternative 5, Gurgaon	3	4	4	2	4	17	68
6	Alternative 6, New Delhi	1	3	3	3	4	14	56
7	Alternative 7 New Delhi	1	5	3	2	3	14	56
8	Alternative 8, New Delhi	2	3	4	2	5	16	64
9	Alternative 9, Gurgaon	2	3	0	4	4	13	52
10	Alternative 10, New Delhi	4	4	4	5	4	21	84

TABLE 5.2 Decision Matrix for Constructability Parameters

S. No.	Criteria Case Study	Database	Category A—Conceptual Planning Stage	Category B—Design Development Stage	Category C—Field Operations Stage	Total Score (Max. 28)	%
1	Alternative 1, Gurgaon	4	7	5	5	21	75
2	Alternative 2, Gurgaon	4	6	5	3	18	64
3	Alternative 3, Gurgaon	5	5	5	3	18	64
4	Alternative 4, New Delhi	5	4	5	4	18	64
5	Alternative 5, Gurgaon	5	6	5	3	19	67
6	Alternative 6, New Delhi	5	5	5	4	19	67
7	Alternative 7, New Delhi	3	4	5	3	15	54
8	Alternative 8 New Delhi	3	3	6	5	17	61
9	Alternative 9, Gurgaon	5	5	6	4	20	71
10	Alternative 10, New Delhi	5	7	6	7	25	89

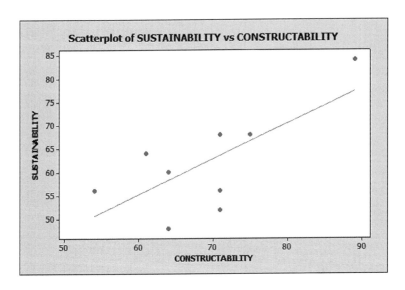

FIGURE 5.1 Positive relationship between constructability and sustainability.

concluded that *the application of Constructability principles shall increase the Sustainability of buildings.*

The study suggests that implementation of constructability practices in projects is likely to have a positive impact on the sustainable development. This construction management tool can act as a simple aid in promoting sustainable development, as the construction industry is facing sustainability challenges.

5.2.2 Task 2—Constructability Practices Followed

A decision matrix was prepared for D&C activities, as discussed in Tables 5.1 and 5.2, and the data were analyzed to observe the percentages of constructability practices followed in construction industry. On close observation of the data collected, the following facts can be assessed and documented regarding various D&C activities, placed under four heads:

5.2.2.1 Database

Database focuses on the project management systems followed by various organizations (in the selected ten alternatives):

1. Project team, excluding the contractor, was formed before the conceptual design stage in 90% of the alternatives.

2. Architect was member of the team in 100% of the alternatives.

3. Project manager was member of the team in 90% of the alternatives.

4. Contractor was not involved, as a team member, before the conceptual design stage in any of the alternatives.

5. Consultants were the team member in 90% of the alternatives.

6. The criteria for selection of team members was experience in similar projects, in 80% of the alternatives.

7. The contract signed was not Design–Build in any of the alternatives.

5.2.2.2 Category A—Conceptual Planning Stage

Conceptual design stage work focuses on the D&C activities management in advance. The decisions taken at this stage have a major role in cost saving. The data collected regarding this category brings forward the following facts:

1. Surveys were conducted in 100% of the alternatives.

2. Proper discussion on construction methods were conducted in 90% of the alternatives.

3. Design/project schedule was prepared in 100% of the alternatives.

4. Ease of field operations were considered in 40% of the alternatives.

5. Discussion on recycling was performed in 40% of the alternatives.

6. Simplification of technical specifications was considered in 90% of the alternatives.

7. Review and implementation of past lessons learned was included in 60% of the alternatives but from the experience and memory of the participants.

5.2.2.3 Category B—Design Development Stage

Design development stage work focuses on various activities related to D&C at the detailing level. The decisions taken at this stage of work are guided by previous stages and have an impact on the field operations. On investigation of the scores obtained by these activities, the following facts can be derived:

1. Only 50% of the alternatives have used advance information technology, during project design development stage.

2. Standardization of design elements was followed in 100% of the alternatives.

3. Review of design by other team members was done in 100% of the alternatives. This team was excluding the contractor in all the alternatives.

4. Considerations were taken for site drainage in 100% of the alternatives.

5. Considerations were taken for water conservation in 100% of the alternatives.

6. Preference was not given to methods for renovation and deconstruction in any of the alternatives.

7. Consideration was taken for the environmentally safe and local construction methods and materials in 80% of the alternatives.

5.2.2.4 Category C—Field Operations Stage
Field operations stage work concentrates on the construction activities at the execution time of the project. On close investigation, the following information was collected:

1. Field task sequencing (CPM/PERT) was adopted in 80% of the alternatives, as bar charts, Gantt charts, or CPM networks.

2. Innovation in available equipment was promoted in 10% of the alternatives only.

3. It was observed that freedom was given to contractors for technical input in 70% of the alternatives. But, in all the cases, they had to take permission from some of the team members.

4. Considerations were taken for adverse weather conditions in 40% of the alternatives.

5. Documentation work of the lessons learned was done in 20% of the alternatives.

6. Waste management on-site was performed during 90% of the alternatives.

7. All the alternatives had regular inspection/meetings on/off-site by consultants.

5.2.3 Task 3—Challenges and Project Management Systems

The study conducted brings forward some of the prevailing challenges faced by team members during various stages of the project. These challenges are documented as causes of delays are presented here. Besides this, there was an observation regarding various types of organization structures or project management systems prevailing in the construction industry. Different systems may have different impact on the functioning of the project.

5.2.3.1 Causes of Delays

Delays in construction projects can occur because of many reasons. Some of the reasons identified during this research are discussed below:

- Delays occur on behalf of contractors when there is shortage of supply of resources, i.e., manpower or supply of material or equipment. This may happen on-site if proper procurement schedules have not been prepared.

- Unexpected delays occur because of adverse weather conditions. Hindrance may occur due to rain, storm, etc. Daily records are maintained on-site for such happenings and hindrance recorded. The contractors are not penalized in such cases.

- Delay occurs because of unavailability of drawings on-site. The resources are idle on-site in such cases and contractor cannot be penalized for such delays.

- Delays may occur because of unforeseen changes in the process of construction or may be the techniques adopted. Such changes are recommended by the contractor but can be adopted only after approval of the project manager, the client, and the consultant. This process may take extra time and delays can occur.

- Delays may occur because the client either changes decision at a later stage of work or delays his decisions. This happens when the client is not involved in decision-making at initial stages of design development or gets influenced by an alternate idea. The change in decision at a later stage introduces a lot of rework and causes delay in the project because the drawings need to be revised.

5.2.3.2 Project Management Systems

The term *team members* refer to the project team formed for the accomplishment of the building project. This team also refers to the participants from the construction industry who are employed for the building D&C. This team includes the client, the architect, the project manager, the consultant(s), and the contractor(s)/constructor at the first level. Later, their subordinates are introduced and appointed on-site to follow the execution of the project. The team may consist of some or all the professionals mentioned depending on the scale, the requirement of the project, or the preference of the client.

Various organizations and systems of project management were examined and analyzed to study the current scenario in the Indian construction industry. The architects, project managers, consultants, and contractors were interviewed personally to understand management systems. On analysis of the data from the working professionals and their experiences, different types of project management systems have been identified.

5.2.3.2.1 Type 1

Figure 5.2 presents Type 1 of the identified project management system. The Client is the foremost and important member as he initiates the project. The Client hires the Project Manager and enters

FIGURE 5.2   Type 1 of project management system.

into contract with him. The Project Manager contracts with the Architect of their choice. The Architect hires the Consultant from his panel based on their own set of criteria. The selected Consultant has to coordinate with the Project Manager. The Contractors for the execution of the project are directly hired by the Project Manager and have to coordinate with the Architect. The situation becomes complicated and a web is formed. The probability of contradiction multiplies. The problems may occur because of Contractor's refusal to certain details and rejection at a later stage of work. The rework is done in such cases and delays are likely to take place.

5.2.3.2.2 Type 2

As shown in Figure 5.3, the Client is the initiator of the project. Client hires the Project Manager and signs a contract with him. The Project Manager then assigns the project to the Architect of his choice. The Client hires the Consultant and the Contractor for the project in consultation with the Project Manager. The Project Manager takes care of all the coordination. The Consultant has to coordinate with the Architect for drawings. The Client depends on Project Manager for all coordination activities including the review meetings.

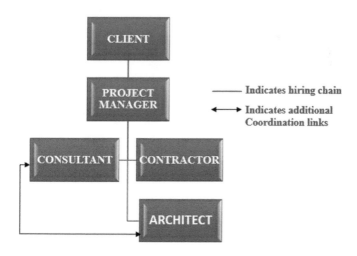

FIGURE 5.3   Type 2 of project management system.

5.2.3.2.3 Type 3

In the third type of system, the Client hires the Architect and gives him all the responsibilities related to the project, as shown in Figure 5.4. The Architect himself is the Project Manager and coordinator in such type of projects. The Architect recommends the Consultants and Contractor for the project to the Client and the contract is signed between the Client and the other team members. The coordination of various activities, review meetings, and any other issues are all dealt by the Architect himself. The work is very close to the Design–Build type of contract. Sometimes the Architect has his team/panel of Consultants and Contractors, who enter into contract with the Client, upon recommendations of the Architect. Such projects are expected to be more successful, have better coordination and reduce the probability of delays, controversies, and ambiguities.

5.2.3.2.4 Type 4

Figure 5.5 shows a system in which the Client hires the Project Manager for entire set of responsibilities related to the project. The Project Manager hires the Architect, Consultants, and Contractor for the Client. The Project Manager is the administrator in this case and takes care of all the coordination among the team members. He hands over the finished product to the Client.

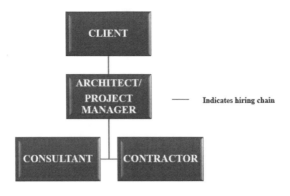

FIGURE 5.4   Type 3 of project management system.

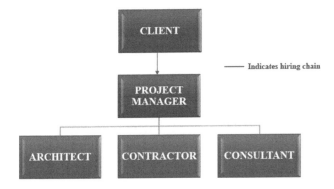

FIGURE 5.5   Type 4 of project management system.

5.2.3.2.5 Type 5

The Client hires the Architect and gives him the responsibility of Project management and coordination also, as shown in Figure 5.6. But sometimes the Client is resistant about the selection of Consultant, so he decides upon the Consultant himself. In such cases, the Architect has to make efforts to coordinate with the Consultant. Whereas the Architect is given freedom to appoint the Contractor from his panel based on his set of criteria. The Architect is the Project Manager in this case, so he has to manage coordination among the team members. If the Client desires to keep the authority of major decisions with himself, he may not allow the Architect to be the administrator.

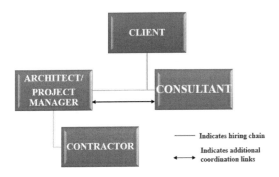

FIGURE 5.6   Type 5 of project management system.

5.3 ANALYSIS OF STUDY 2

This section focuses on the study of interrelationship between various D&C activities in terms of dependency and independency. The building construction exercise is a combination of D&C activities having linkages with each other. The inventory of such activities could be very long. The important activities related to D&C, which have major role to play and influence the constructability, are identified and further used in the study. An elaborate list of thirty activities is prepared in the three categories.

The placement of an activity in a category is established through pilot study conducted and content validity test, which justified the parameters of constructability to be placed in a certain category.

Initially the responses are categorized based on frequency. The responses collected for each activity are divided into four categories:

- Least dependent—(0–7)

- Partially dependent—(8–15)

- Moderately dependent—(16–22)

- Highly dependent—(23–30)

The least dependent activities, i.e., scoring below 7 out of 30 responses, are neglected in the formulation of Design Structure Matrix (DSM). Rest of the relationships having partial, moderate, and high dependency are considered.

5.3.1 Design Structure Matrix

DSM is used for study of interrelationship between activities. A DSM is defined as a compact, matrix representation of a project network. It provides an idea about various activities of any process that are interrelated, what information is required to start an activity, what activity will be followed by any previous activity, manages complex projects, does task sequencing and iterations (Yassine, 2004). Steward (1981) also defined DSM as "a methodology to handle dependencies and relations between activities." Browning (2001) explains that DSM is a representation and analysis tool for the system modeling. A DSM displays the relationships between components of a system in a compact, visual, and analytically advantageous format.

A DSM is also known as "Dependency Structure Matrix" (Danilovic and Browning, 2007). However, there are other terms also like the Problem-Solving Matrix and Design Precedence Matrix. It is a management tool applied in project management. It provides a project representation that allows for feedback and cyclic task dependencies.

This study is focused on activity-based DSM. These are used for modeling processes and activity networks based on the activities, their information flow, and other dependencies.

A DSM is thus prepared for each of the categories A, B, and C on the same guidelines. The DSM is worked out by partitioning and tearing of Principal circuits. The final DSM obtained after iterations is analyzed for dependency of activities. The flowcharts showing network before and after application of iterations are prepared to highlight the significant changes and benefits of utilizing DSM for project management.

The process of solving DSM can be explained as follows: The original DSM has activities above and below the diagonal. The marks above the diagonal are feedback marks, which should be removed to make the project efficient. The first step is to partition and then apply tearing to the longest circuit, to break the BLOCK. The matrix rearranges itself after tearing the circuit and a new matrix is formed after partitioning. The iterations are performed until there appears no 0 mark above the diagonal. Sometimes, a number appears for activities above the diagonal, which also means that the matrix is solved. All the 0 marks are shifted below the diagonal. Hence, the final matrix is obtained.

5.3.2 DSM for Category A—Conceptual Planning Stage

Table 5.3 represents the dependent and independent activities, which are finally used for preparing the DSM.

DSM was prepared for the data collected in Table 5.3. The original DSM for activities in category A—Conceptual planning stage is shown in Figure 5.7.

TABLE 5.3 Dependency of Activities in Category A—Conceptual Planning Stage

S. No.	Activity	Depends on the Activity
1	Selection of architect	Independent
2	Shortlisting of contractors	1, 5
3	Discussion on how to make construction process easier	1, 2, 5, 10
4	Conducting surveys	Independent
5	Selection of construction methods	1, 2, 3, 7
6	Working out construction schedule	1, 2, 3, 5
7	Laying out site efficiently	1, 4, 5
8	Discussion on recycling	1, 2, 3
9	Simplification of technical specifications	1, 2, 5
10	Review and implementation of past lessons learned	Independent

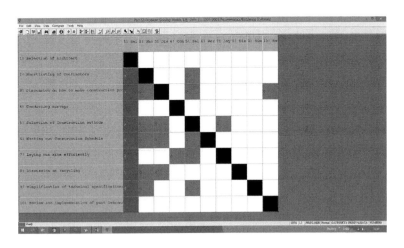

FIGURE 5.7 Original DSM for category A—Conceptual planning stage.

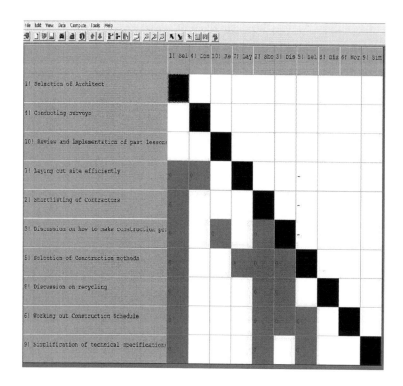

FIGURE 5.8 Final DSM for category A—Conceptual planning stage.

The iterations were performed, and the final DSM obtained, as shown in Figure 5.8.

The final DSM has activities which are independent and dependent. These can be planned on the basis of DSM, as worked out here. These can also be used for preparing CPM networks for the activities. The activities that have been identified by DSM as independent activities are: selection of architect, conducting surveys, and review and implementation of past lessons learned. The final DSM is further analyzed to find out the dependency of activities.

- Laying out site efficiently depends on selection of architect and conducting surveys.

- Shortlisting of contractors depends on selection of architect.

- Discussion on how to make construction process easier depends on selection of architect, shortlisting of contractors, and review and implementation of past lessons learned.

- Selection of construction methods depends on selection of architect, shortlisting of contractors, discussion on how to make construction process easier, and laying out site efficiently.

- Discussion on recycling depends on selection of construction methods, selection of architect, shortlisting of contractors, and discussion on how to make construction process easier.

- Working out construction schedule depends on selection of architect, shortlisting of contractors, discussion on how to make construction process easier, selection of construction methods, and laying out site efficiently.

- Simplification of technical specifications depends on selection of architect, shortlisting of contractors, and selection of construction methods.

The iterations done in DSM lead to a simplified network of activities. The results thus obtained are utilized for making simple networks that can be used as guide for construction activities, in sequencing the different jobs. The network for Category A before the application of DSM is shown in Figure 5.9.

As shown in Figure 5.10, the network after the application of DSM sorts out the activities and gives a simplified logical sequence of activities to be followed.

Selection of architect, conducting surveys, and review and implementation of past lessons learned are identified as independent activities. These activities can be conducted without waiting for other activities to initiate. Laying out site efficiently depends on selection of architect and surveys conducted. The architect is involved in shortlisting of contractors and is responsible for

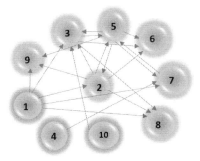

FIGURE 5.9   Original network for category A—Conceptual planning stage, before the application of DSM.

FIGURE 5.10   Simplified network for category A—Conceptual planning stage, after the application of DSM.

making discussions on how to make construction process easier, selecting the construction methods, selection of recycled materials, working out schedules, and simplification of technical specification. The selection of contractors shall lead to selection of construction methods and their suggestions can be incorporated for making construction process easier, recycling and specification selection. They are also involved in preparation of construction schedules. Review and implementation of past lessons learned helps in increasing the ease of construction because the past mistakes are not repeated. Discussion on how to make construction process easier would help in selection of appropriate construction methods, recycling practices, and affect the construction schedule. Laying out site efficiently shall guide way for selection of appropriate construction methods to be adopted and in working out the construction schedules.

5.3.3 DSM for Category B—Design Development Stage

Table 5.4 represents the dependent and independent activities, which are finally used for preparing the DSM.

A similar analysis of data using DSM is performed and dependency of activities is analyzed. DSM is prepared for the data collected in Table 5.4. The original DSM for activities in category B —Design development stage is shown in Figure 5.11.

The iterations are performed, and the final DSM obtained, as shown in Figure 5.12.

The final DSM has activities which are independent and dependent. These can be planned accordingly. The activities which have

TABLE 5.4   Dependency of Activities in Category B—Design Development Stage

S. No	Activity	Depends on the Activity
1	Development of design and procurement schedule	Independent
2	Selection of contractor	Independent
3	Selection of subcontractors/vendors	2
4	Use of advance information technology	Independent
5	Standardization of design elements	10, 11
6	Review of design by other team members	Independent
7	Small scale physical models/3D drawings	Independent
8	Considerations for site drainage	Independent
9	Considerations for water conservation	Independent
10	Concern to reduce scaffolding	1, 2, 5
11	Preference of methods/materials for renovation and deconstruction	6
12	Consideration for environmentally safe materials/methods of construction	2, 6

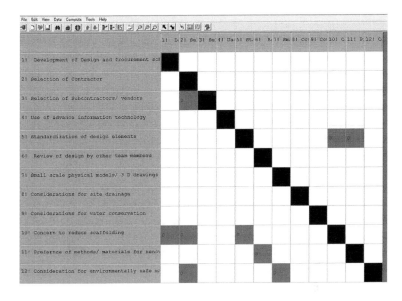

FIGURE 5.11   Original DSM for category B—Design development stage.

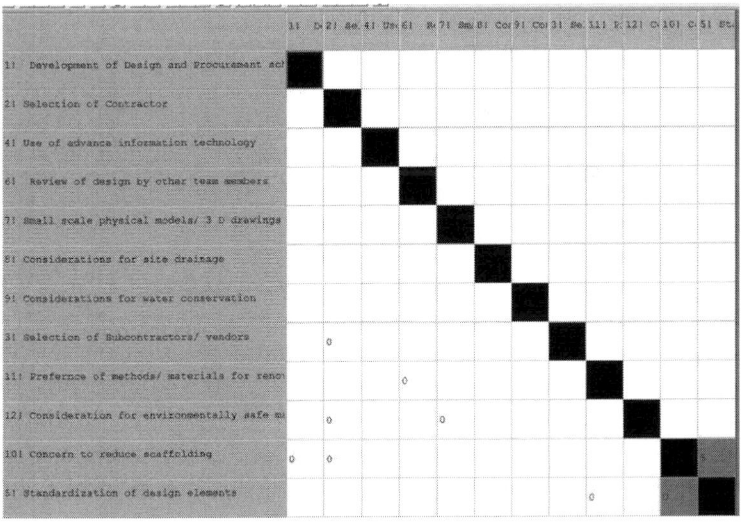

FIGURE 5.12   Final DSM for category B—Design development Stage.

been identified by DSM as independent activities are: development of design and procurement schedule, selection of contractor, use of advance information technology, review of design by other team members, small-scale physical models/3D drawings, considerations for site drainage, and considerations for water conservation. The dependent activities are as follows:

- Selection of subcontractors/vendors depends on selection of contractor.

- Standardization of design elements depends on concern to reduce scaffolding and preference of methods/materials for renovation and deconstruction.

- Concern to reduce scaffolding depends on development of design and procurement schedule and selection of contractor.

- Preference of methods/materials for renovation and deconstruction depends on review of design by other team members.

- Consideration for environmentally safe materials/methods of construction depends on selection of contractor and small-scale physical models/3D drawings.

The network for Category B before the application of DSM is shown in Figure 5.13.

As shown in Figure 5.14, the network after the application of DSM sorts out the activities and gives a simplified logical sequence of activities to be followed.

Development of design and procurement schedule, selection of contractor, use of advance information technology, review of design by other team members, small-scale physical models/3D drawings, considerations for site drainage, and considerations for water conservation can be initiated earlier and simultaneously, by the project team members. The selected contractor further collaborates and makes the team with sub-contractors/vendors.

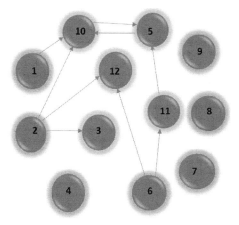

FIGURE 5.13 Original network for category B—Design development stage, before the application of DSM.

The planners concern to reduce scaffolding helps in standardization of design elements. It is in the purview of the team to prefer methods/materials for renovation and deconstruction, which in turn depends on the standardization of the design elements. The two are closely related to each other. Development of design and procurement schedule is related to time management which shall also be governed by reduced scaffolding. The selection of the contractor shall lead way for the concern to reduce scaffolding and selection of environmentally safe materials/methods of construction. If the contractor is concerned for all these factors, then only it would be possible to save environment from the pollution and disturbance caused during the construction project. Review of design by other team members brings forward ideas related to methods/materials for renovation and deconstruction, because of the expertise of the team members in their areas of specialization. Different stages of work and environmentally safe materials and methods of construction can be planned, and phases of construction decided which shall help in working on-site, without causing much disturbance to the neighbors.

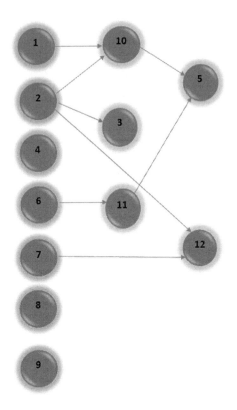

FIGURE 5.14 Simplified network for category B—Design development stage, after the application of DSM.

5.3.4 DSM for Category C—Field Operations Stage

Table 5.5 represents the dependent and independent activities, which are finally used for preparing the DSM.

The original DSM for activities in category C—Field operations stage is shown in Figure 5.15.

The iterations are performed, and the final DSM obtained, as shown in Figure 5.16.

After the analysis of data, the final DSM has activities which are independent and dependent. These can be planned accordingly. The activities which have been identified by DSM as independent activities are: field task sequencing (CPM, etc.), freedom

TABLE 5.5 Dependency of Activities in Category C—Field Operations Stage

S. No.	Activity	Depends on the Activity
1	Field task sequencing (CPM) etc.	Independent
2	Use of temporary material/system on-site	3, 7
3	Innovation in available equipment	4, 7
4	Freedom to contractors for technical input to improve the construction process	Independent
5	Use of pre assembly in case of adverse weather	2, 4
6	Documentation work of the lessons learned during the project execution stage	Independent
7	Waste management on-site	2, 4, 8
8	Regular inspection/meetings on/off-site by consultants	Independent

to contractors for technical input to improve the construction process, documentation work of the lessons learned during the project execution stage, and regular inspection/meetings on/off-site by consultants.

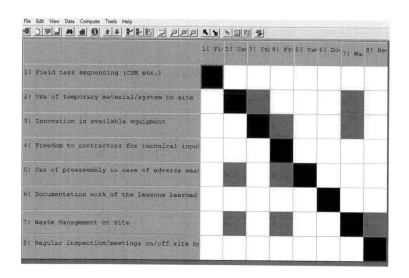

FIGURE 5.15 Original DSM for category C—Field operations stage.

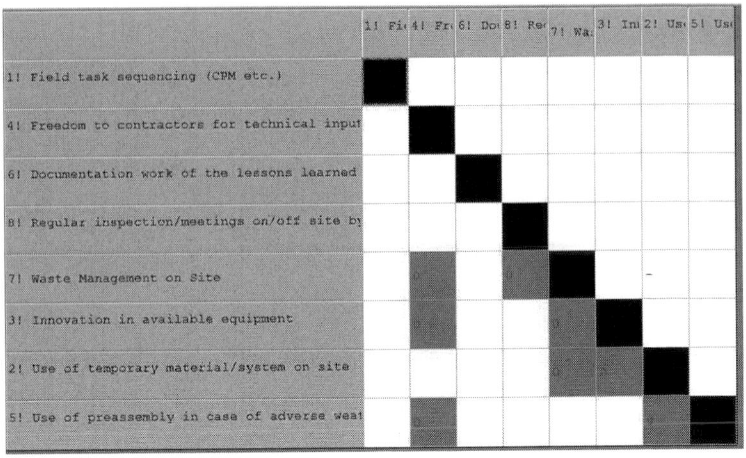

FIGURE 5.16   Final DSM for category C—Field operations stage.

- Waste management on-site depends on regular inspection/ meetings on/off-site by consultants, and freedom to contractors for technical input to improve the construction process.

- Innovation in available equipment depends on freedom to contractors for technical input to improve the construction process, and waste management on-site.

- Use of temporary material/system on-site depends on innovation in available equipment and waste management on-site.

- Use of pre assembly in case of adverse weather depends on use of temporary material/system on-site, and freedom to contractors for technical input to improve the construction process.

The network for category C before the application of DSM is shown in Figure 5.17.

As shown in Figure 5.18, the network after the application of DSM sorts out the activities and gives a simplified logical sequence of activities to be followed.

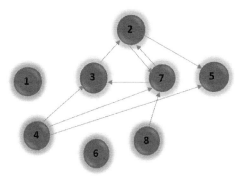

FIGURE 5.17 Original network for category C—Field operations stage, before the application of DSM.

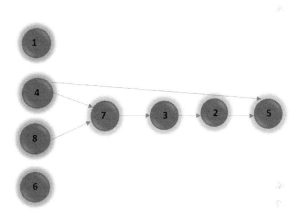

FIGURE 5.18 Simplified network for category C—Field operations stage, after the application of DSM.

Field task sequencing (CPM, etc.) is to be done at initial stage of work. Freedom to contractors for technical input to improve the construction process, documentation work of the lessons learned during the project execution stage, and regular inspection/meetings on/off-site by Consultants can take place independently during the field operations stage of work. Regular inspection/meetings on/off-site by consultants help in controlling and managing waste on-site. This helps in regular monitoring of the work and smooth working of the project. If the Contractors are given freedom for technical input to improve the construction process, it can result in the innovation in available equipment for better working and saving time also, like if they use preassembly in case of adverse weather conditions. Waste management on-site can also lead to innovation in available equipment to reduce and segregate or manage waste and use of temporary material/system on-site.

REFERENCES

Browning, T. R. (2001). Modelling the customer value of product development processes. (Proceedings of International Symposium, INCOSE), Melbourne, Australia, 11(1) 592–599. DOI:10.1002/j.2334–5837.2001.tb02346.x.

Danilovic, M., & Browning, T. R. (2007). Managing complex product development projects with design structure matrices and domain mapping matrices. *International Journal of Project Management, 25*(3), 300–314. DOI:10.1016/j.ijproman.2006.11.003.

Muller, D. G. (2002). *Sustainable Architecture and Urbanism: Concepts, Technology, Example*. Basel: Birkhuser.

Singh, J. (2008). *Methodology and Techniques of Social Research*. New Delhi: Kanishka.

Steward, D. V. (1981). The design structure system: A method for managing the design of complex systems. *IEEE Transactions on Engineering Management, 28*, 71–74. DOI:10.1109/TEM.1981.6448589.

TERI. (2004). *Sustainable Building, Design Manual, Vol. 1*. New Delhi: TERI.

Yassine, A. A. (2004). An introduction to modelling and analyzing complex product development processes using the design structure matrix (DSM) method. *Journal Urbana, 51*(9), 1–17.

CHAPTER 6

General Recommendations

One of the major issues identified in the construction industry is lack of coordination among team members. The team is not formulated completely at the initial stage of design. A team shall ideally comprise representative from all the fields: the Client, the Architect, the Project Manager, the Consultant, and the Contractor. Unfortunately, it was observed that the team formed at the initial stage of design does not include the contractor, in majority of cases. Quite a lot of architects agreed to the fact that they encountered problems during the project execution stage because the contractor refused to agree to some of the details at a later stage of work. This resulted in a lot of rework and drawings had to be reproduced, thus wasting resources and the project was delayed, at times. Generally, the contractors are selected on the lowest bid and at a later stage of work. In some cases, they encounter problems with the fixing detail of the specifications because of lack of experience in that material. The Design–Build contracts are theoretically considered as the best type of contracts, but unfortunately were not witnessed in any of the case studies conducted during the research. The barriers toward inclusion of Contractor in the team at the initial stage of design have been identified as follows:

- Increase in specialization and complexity of projects has made Design–Build projects less popular.

- The bid for contract is generally based on Bill of quantities. This makes it an obvious reason for the Contractor to enter the team, only after all the drawings are prepared.

- The most preferred contracts signed between the Contractor and the Client are item rate contracts, which can be signed only after the drawings are prepared.

- The charges/fees of the Contractors initiates when they are introduced in the project.

- The Client is not interested in bearing expenses of Contractors, when the design is at initial stage.

- The involvement of Contractors at the initial stage of design is generally not entertained by Architects, as there is a belief that the Contractor's suggestions are biased toward ease of construction and obstruct in creativity and design development by the Architect.

- The practice of excluding Contractor from team at the initial stage of design has been generalized.

Some architects agreed to the fact that the involvement of contractors at an early stage of design is beneficial because it saves time, which may otherwise be invested in reworks at a later stage due to non availability of any of the resources. The suggestions from the contractors help them in choosing the appropriate materials and techniques for specification writing. This helps in selecting the right material and right workmanship depending upon their availability, feasibility, and maintenance experience. The exposure of contractors to field experience and knowledge about forthcoming issues is beneficial as it saves time, energy, and money that could otherwise be wasted in redrawing the details when the problem is encountered. The traditional bidding process continues, although the benefits of inclusion of contractor at early stage are realized. A fruitful suggestion during the discussion with the Project Managers and Architects was that instead of bidding with the contractors at the initial stage of design, an alternate could be adopted, and shortlisting of contractors can be done. The selection of a candidate in the panel shall depends on the nature of project and the experience of contractor in a similar scale of project. They shall be invited for timely suggestions as and when required, and each one of them shall have an equal chance of winning the bid. This methodology can bring twofold benefits. First, variety of suggestions would be wide and exclusive, which would be profitable for the project. Second, the contractors would be keen on bidding, and more appropriate and logical bids would be prepared because of better understanding of the project in advance. The interaction of shortlisted contractors with the Client would provide him a fair idea about their knowledge and experiences. This would be helpful during bidding process and guide in taking legitimate decision.

Some of the possible solutions can be listed as follows:

- Formation of panel of contractors, depending on the nature of project.

- Identifying panel members at the initial stage of design.

- No discussion on form and function of the building with the panel.

- The discussion shall include alternative specifications, available resources, increasing site efficiency options, waste management issues, recycling issues, and so on.

- Document the discussions for future use.

- Review past experiences learned and incorporate them.

The complexity of construction projects has introduced disintegration in the construction projects. The team members are confined to their specific areas of interest, thus leading to lack of coordination in the team. Team coordination needs to be strengthened, at priority.

Some of the possible solutions for increasing coordination can be listed as follows:

- Formulation of team in the beginning of project.

- The educational qualification of team members can be one of the criteria for selection of team participants, to maintain the equity of thoughts and communication.

- Initiative for making team should be taken by the Client or Architect.

- A team should comprise of a Client, an Architect, a Project Manager, a Consultant, and a Contractor.

- Documentation/records of important decisions, duties, and progress should be maintained.

- All participants should be given fair opportunity to express their ideas and open discussions should be carried out in friendly atmosphere, in the interest of the project.

It was agreed by the practitioners of the construction industry that the projects witness better coordination and run more smoothly when the Architect is the Project Manager and administrator of the project team.

There is an emergent need to stress upon the importance of project management practices related to various design and construction activities based on constructability and increase the awareness level of architects. The academicians can play a very important role in propagating and creating awareness regarding the issues concerned with D&C practices, as they are the playing a key role in building up of budding architects. The academic institutions and government organization can contribute by organizing orientation programs for promoting the information regarding issues related to various D&C activities based on constructability, which can enhance the project quality and promote sustainable development.

It is also very important to highlight the importance of constructability practices related to various D&C activities and increase the awareness level of practicing and non practicing architects. The academicians must be involved in consultancy and professional practice for promotion of best D&C practices. The academic institutions should support their staff and promote professional practice so that various D&C activities based on constructability can be displayed at a global level for sustainable development.

CHAPTER 7

Project Management System Models

7.1 INTRODUCTION

Good administration is the key to success of the project. This section focuses on the administrative issues during various stages of project. Integration of team members at initial stage is not helpful without good coordination. It requires the coordination of highest degree among the team members to achieve the profitable end results. The interactive sessions with various stakeholders are concluded in the form of suggested models regarding administration of the project during different phases of the project. Models are worked out and proposed here for coordinated management of the project.

7.2 THREE STAGES FOR MANAGING PROJECTS

The professionals in the construction industry presented their concern regarding the status and governance for various participants, at various stages of project. Accordingly, the responsibilities and leadership should keep rotating among the team members during different stages of work. Stagnation of authority with one team member should be avoided and the administrative powers should be in rotation. The professionals suggested that different types of decisions are to be taken at different times of a project. There should be one administrator and all other team members should report to him, for better coordination. This study identifies three stages for administering the projects, based on discussions with the professionals. These stages are listed as follows: *decision stage*, *design stage*, and *field operations stage*. The flow of information is suggested for the three stages, as shown in Figures 7.1–7.3.

Different *models* are suggested based on these recommendations, for the project management, at different stages of a project. The salient features of the suggested *model* are as follows:

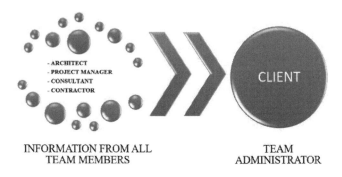

- ARCHITECT
- PROJECT MANAGER
- CONSULTANT
- CONTRACTOR

CLIENT

INFORMATION FROM ALL
TEAM MEMBERS

TEAM
ADMINISTRATOR

FIGURE 7.1   Information flow for decision stage.

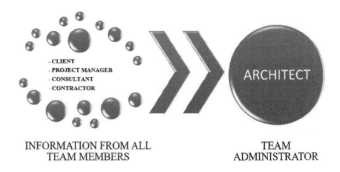

- CLIENT
- PROJECT MANAGER
- CONSULTANT
- CONTRACTOR

ARCHITECT

INFORMATION FROM ALL
TEAM MEMBERS

TEAM
ADMINISTRATOR

FIGURE 7.2   Information flow for design stage.

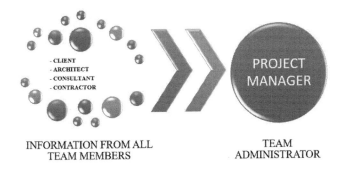

- CLIENT
- ARCHITECT
- CONSULTANT
- CONTRACTOR

PROJECT MANAGER

INFORMATION FROM ALL
TEAM MEMBERS

TEAM
ADMINISTRATOR

FIGURE 7.3   Information flow for field operations stage.

- Client will bring together the project participants: Architect, Project Manager, Consultant, and Contractor.

- Team will be integrated at the initial stage.

- Information, review meetings, and discussions will take place.

- The information will flow to team administrator, i.e.,

  - To the Client at the decision stage

  - To the Architect at the design stage

  - To the Project Manager at the execution stage

- If the team administrator agrees to a decision, the action is taken.

- If the team administrator disagrees to a decision, the discussion is open for further improvement and analysis.

7.2.1 Model for Decision Stage

It is the first stage of the project. Here all the major introductory decisions are taken by the Client. The Client provides the major resources for the project. He hires the Project Management Consultancy or Architect and all the other team members. The Client is the decision maker at this stage, so it is important that

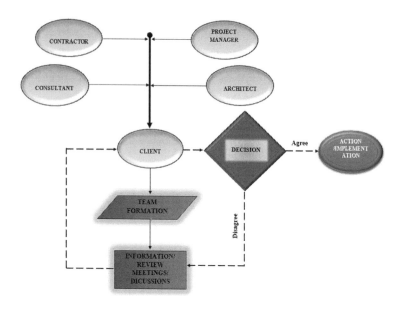

FIGURE 7.4   Model for decision stage.

all the information flow to him. Figure 7.4 shows the Client as the administrator for this stage and all information flows to him.

7.2.2 Model for Design Stage

It is the next stage of the project that includes both conceptual planning and detailed design development. Architect is the most important participant at this stage; all the success and fame of the project is on his shoulders. He must provide the basic plan, invite input, coordinate team, get approvals, prepare estimates, and provide all the necessary working drawings for the execution of the project on the site. At this stage, the information from all the resources should flow to Architect and he should be the administrator, as shown in Figure 7.5.

7.2.3 Model for Field Operations Stage

It is the final stage of the project, when the actual construction work on the site starts. At this stage of project, the most important

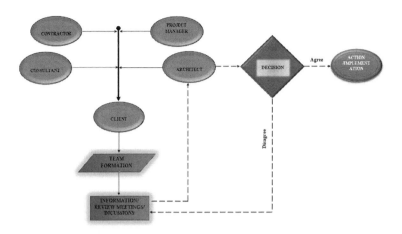

FIGURE 7.5   Model for design stage.

member of the team is the Project Manager, as he must coordinate between different team members and get the work executed on site. All the significant information should reach the Project Manager for the success of the project and he should be the administrator at this stage of work, as shown in Figure 7.6.

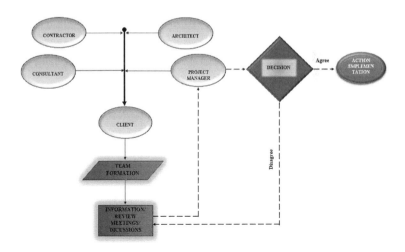

FIGURE 7.6   Model for field operations stage.

CHAPTER 8

Responsibilities and Role of Project Participants

8.1 INTRODUCTION

In novel times, there has been a gap in the construction industry regarding roles played by different professionals. The division of responsibilities in projects has laid foundation for problems of varying degree. Every participant has their areas of specialization and they have confined themselves to their specific fields. They do not interfere in another team member's domain. The work is distributed and each one is bothered only about their share of responsibility. The overall interest of the project may suffer in such cases and the probability of problems due to lack of coordination increase. A study of the roles played by different key players in the construction industry is conducted. Their responsibilities and roles in the present scenario have been identified and discussed here. However, their roles will depend on the type of project management system adopted.

8.2 RESPONSIBILITIES AND ROLES OF PARTICIPANTS

8.2.1 Role of Client

The Client plays a crucial role in the design and construction process because he is the one who initiates the project. All the major financial controls and decisions are taken by the Client and depend on his wisdom. The Client provides the most important resources for the project: the land and the finance. Some of the important contracts are signed by the Client himself. The Project Manager and the Contractor are appointed by him and he enters in contract with them directly in most of the cases. Sometimes the responsibilities are shared by the Project Manager or an Architect appointed by the Client himself. It has been observed, during the study of various projects and interviews with resource persons, that Contractors are not involved in the projects at an early stage of design. The suggestions and pragmatic ideas of the Contractor can be of great help in reducing unforeseen problems during the execution of the project. The Client should play a key role in promoting an integrated comprehensive team formation at the conceptual planning stage of design, so that the input from the experience of all the team members can be utilized and incorporated in the design development stage of the project.

The Client is generally interested in saving money and early completion of the project, so that the returns and services can be taken from the finished project as early as possible. Sometimes, in his celerity, some of the important issues related to environment are compromised. The Architect contributes his finest services and ideas, which may or may not be consented. The Contractors may not be allowed extra time and facilities to take care of measures and methods of environmentally safe construction on site. The lack of awareness of the Client and resistance may lead to great environmental hazards, which could otherwise be avoided. It is a matter of great concern and duty of the Client to accept and promote the constructive and optimistic ideas and suggestions of the Architects and other team members. The Client can play a crucial role in saving the environment, by investigating and being vigilant on some of the issues such as

- Giving preference to Architects working on principles of constructability and environmental issues.

- Encouraging tenders from Contractors, with alternative solutions of safe construction methods/techniques related to environmental issues.

- Introducing concessions for D&C practices that take care of constructability and environmental issues.

- Giving due concern and relaxation in time that may be increased because of the implementation of environment friendly methods and techniques of D&C practices.

- Promoting and supporting the use of recycled materials and renewable resources of energy.

- Introducing the stakeholders at an early stage of the project so that their experiences and suggestions can be timely introduced in the project and maximum benefit achieved.

8.2.2 Role of Architect

The Architects are key role players in the project team because their design is to be executed on-site. Their wisdom, knowledge, and experience lead to creation of wonders. The Architects are hired either by the Client himself or by the Project Manager appointed by the Client. The criteria for selection of an Architect as observed can be listed as follows:

- The personal contacts of Client or the Project Manager.

- The reputation of an Architect's creative abilities.

- The specialization of an Architect in projects of similar nature.

- The Architect has handled a project of a similar scale/ budget/turnover.

- The fascination of the Client for a style followed by the Architect.

- The expertise of an Architect in a building material.

- The experience of an Architect in professional practice, in number of years.

- The accessibility and approach of an Architect.

- The competition winning Architect is awarded the project.

- The Architect may be panelled with some organization/ government bodies.

- The Architect is recommended by the end users of his previous projects.

The Architect prepares the design proposal and gets the approval of the Client. He provides the necessary drawings for the execution of the project. In case Project management consultancy is not hired by the Client, the Architect is the Project Manager for the project and is involved in all the major decisions related to the project. The Consultants and Contractors are hired by the Client on his recommendation. The Architect is the administrator and he coordinates with all other team members. Such projects work more effectively, and probability of delays and lack of coordination are minimized. In some cases, even if the PMC is hired, the Architect is the administrator of the project as per the directions of the Client. The Architect can play important role by focusing on following issues:

- Insist on early involvement of the Contractor in the project.

- Explain the idea of constructability and its benefits to the team members.

- Break off the psychological barrier that the involvement of Contractor shall hinder his creativity.

- Open discussions and invite suggestions from team members.

The Architect must design the building keeping in mind several issues, especially those related to the environment. They should convince the Client for early team formation, and energy-saving materials and methods/techniques.

8.2.3 Role of a Project Manager

The Project Manager is introduced in the project by the Client at an early stage. He signs a contract with the Client and shares the responsibilities of the project coordination. There are two ways of assigning duties to the Project Manager, which are discussed below:

- The PMC signs a simple contract agreement with the Client. It provides the staff for working on the project and the Client agrees to pay salary to the staff. This salary is fixed by the PMC on a monthly basis. The expenses of the head office are also born by the Client. According to the PMC, this type of contract is safe and risk free, as the PMC does not bear any penalty for delays and unforeseen hurdles.

- The PMC signs a contract with the Client as total project package service provider. Part payments are done by the Client, on a timely basis. According to the PMC, this type of contracts is full of risk and is not generally preferred.

In the majority of the cases (where a PMC is involved in the team), the Project Manager is first appointed by the Client. The Architect, Consultant, and Contractor are hired by the Client in consultation with the Project Manager. The Project Manager is the assigned administrator of the project and heads all the review meetings. He has many duties to perform:

- The Project Manager organizes meetings on a regular basis to monitor the progress and quality of work.

- The Project Manager is responsible for coordination among the team members to avoid any ambiguity.

- The Project Manager also takes care of lawsuits, if any.

- The Project Manager recommends other team members to the Client, for inclusion in the project.

8.2.4 Role of a Contractor

In the present scenario, the firms generally call Contractors for bidding only after the design stage is complete. The Bill of quantities is prepared and an estimate of approximate cost of the project is prepared. This methodology leaves less scope for the Contractor's involvement and contribution to the project in terms of sharing from his experience. The selection criteria are different, and they are adopted as per the choice and suitability of the project participants. Some of these criteria can be listed as follows:

- The turnover of his previous projects is appreciable.

- The Contractor has experience with projects of a similar nature.

- The Contractor has specialization in a particular type of work. Workmanship is also available for some specific technical work.

- The Contractor's previous project's budget was of the same quantum.

The Client has to pay the Contractor, so he has the authority to choose the Contractor. Generally, the Client selects the Contractor on recommendations of the Project Manager or the Architect, after some negotiations. The Contractor can play an important role in the following ways:

- He should be concerned about the environmental issues also and convince the Client/Project Manager at the initial stage about their implementation. The construction practices should not bring any damage to the environment.

- He should convince the team members regarding innovations possible in the construction methods and technology to gain maximum benefits for the project.

- He should break off the psychological barrier that they are not accepted as part of team, at the initial stage of the project.

8.2.5 Role of a Project Engineer

The Project Engineer/Site Engineer is appointed by the PMC/Architect, respectively, on the site to carry out the responsibilities on their behalf. They are the representatives of the PMC or Architect. Their primary duty is to maintain correspondence with their officials and take care of construction work on-site. They are also responsible for coordination on-site and report to their officials, in case of any ambiguity or misunderstanding.

8.2.6 Role of Financial Institutions

The financial bodies can play an important role in the promotion of constructability features and help in sustainable development. Their aids can enhance the quality of life and help reduce environmental degradation to a great extent. These institutions are in commendable capacity and can serve the society well by providing necessary grants. Some of the important initiatives have been listed below, which can be exercised and explored further.

- Buildings promoting and supporting constructability principles in their working can be given incentives.

- Additional funding can be provided for buildings which are designed to work on energy saving and increased efficiency parameters.

- Low interest rates can be provided exceptionally, for buildings managed on constructability issues and designed on sustainable features. Such features can be recognized, enlisted, and approved by the development authorities.

8.2.7 Role of Regulatory Bodies

The regulatory bodies are the backbone of development. They are empowered to set rules and regulations for the sustainable and safe development of the society. It is imperative that this body keeps a check on the type of D&C practices and acts for promoting them for sustainable development. Some of the important initiatives that are exercised and can be explored further have been listed below:

- Certain features and practices of constructability can be recognized, enlisted, and approved by the development authorities as mandatory for buildings. Such features may vary according to the typology and scale of the building.

- Buildings promoting and supporting constructability parameters can be given some exemptions in terms of Floor area ratio or ground coverage, etc.

- Knowledge centers can be established for promotion and counseling of participants from the construction industry to learn best D&C practices.

- Penalties can be introduced for noncompliance of the laws related to energy saving.

- Media could be utilized for spreading the useful information related to best D&C practices.

- The government organizations can promote and set examples for others by constructing such buildings for themselves.

- Landfill taxes may be imposed by local authorities to reduce the waste generation and promote waste segregation, reuse, and recycling of building materials.

- Subsidies can be provided on the recycled products for their promotion in the construction industry.

CHAPTER 9

Checklist for Promotion of Constructability

9.1 INTRODUCTION

Checklists are prepared for various stages of the project after the studies conducted. These are summarized in three categories: conceptual planning stage, design development stage, and field operations stage. The concern for these issues shall lead to tangible and intangible achievements, after completion of the project.

9.2 GENERAL DESIGN AND CONSTRUCTION PRACTICES

1. The *project team should comprise Client, Architect, Project Manager, Contractor (construction personnel), and Consultants*, and should be formed before the conceptual design stage. (The input from the experiences of all team members can be incorporated at initial stage of work. This helps in reducing rework at a later stage.)

2. Criterion for selection of team members should be *experience in similar projects*, rather than the lowest bid. It is preferable to *adopt a Design–Build contract*. (This shall minimize the problems, which occur due to lack of coordination. It was observed in some cases that the Architect was given freedom to choose members of the team. In such cases, the coordination among participants is experienced as smooth by many researchers.)

3. *Proper surveys* including the site survey for topography and vegetation available, local building material survey, etc., should be conducted before initiating the project work.

4. There should be a *discussion on selection of appropriate construction methods* at the initial stage of the design. *Ease of field operations* should be discussed among the team members. (These considerations can be regarding the movement of construction personnel or equipment or may be maintenance-related work on-site, while the construction work is going on.)

5. *Design/project schedule* should be prepared at the conceptual planning stage. Use *advance information technology* for design development. Management tools such as *field task sequencing* should be done with Critical Path Method/ Project Evaluation and Review Technique to regulate the construction work.

6. There should be a *discussion on designing for recycling* and adopting recycled materials for the project. *Simplification of technical specifications* should be well considered, depending upon the workmanship and building material availability.

7. *Review and implementation of past lessons learned* should be done. (This helps in minimizing errors and saving time. The experience of all the team members, in their expertise

areas, can be of great help.) At the design development stage of work, *review of design should be done by other team members also.*

8. *Standardization of design elements* should be encouraged to make the project economically more viable.

9. Proper considerations should be adopted for *site drainage* at all stages of work as well as for operation of the project. Concern and considerations are recommended for *water conservation* on-site. *Waste management* on-site is important at all stages of the work. (During execution special measures should be adopted for managing waste by either reutilizing on-site itself or disposing after proper segregation under the laws.)

10. Detail procedure should be worked out in advance at the planning stage for the *renovation* during maintenance life of the building, as well as *deconstruction* after the useful life of the building is over.

11. The design should consider for *environmentally safe* and *local construction methods and materials.* (This shall help in reducing the embodied energy and attaining proper workmanship.)

12. *Regular review meetings* should be conducted during the field operations stage. The meetings should be headed by an administrator (decided by the Client in most of the cases). It would be the responsibility of the administrator to collect the information and draft the responsibilities of each participant, in advance. *Inspections*/meetings on/off-site by consultants help ensure proper working of the project. (Such meetings should focus on the architectural design, working drawing clarifications and understanding, project schedule, issues of technical specification, if any, and hurdles in timely delivery of the project.)

13. *Smooth communication* among the participants involved in the project can be managed by regular review meetings, giving freedom to Architect for choosing the participants, coordination by Project Manager, or coordination by the client himself.

14. Considerations should be taken for *adverse weather conditions.* (Some measures in design can be introduced keeping in mind the forthcoming seasons, e.g., preference for prefabrication, etc. This helps in saving material and manpower as well as reduces the chances of project delay.)

15. The contractors should be encouraged to make *innovation/ modification in available equipment,* to enhance the quality and speedup execution work of the project, and to save energy. *Freedom* should be given to contractors for sharing technical input at all stages of work. (This helps in economic savings as well as smooth execution of the project.)

16. Proper *documentation work of lessons learned,* at all stages of the work, should be done to retain as feedback from the project. These can be used as do's and don'ts for future projects and hence save time and energy that would otherwise be wasted in repeating the errors.

9.3 CHECKLIST FOR THREE STAGES OF A PROJECT

After the study of the various aspects of constructability and its crucial role in managing projects, the following checklist has been formulated. Thirty activities are classified into three different categories. It would be beneficial to check them and make projects more efficient by introducing the concept of constructability.

9.3.1 Checklist for Conceptual Planning Stage

In this stage, the client identifies the site and tests the feasibility of the project by doing elementary work in estimate, plans, etc. Thereafter, he decides of whether to proceed with the project.

1. The program for constructability is discussed and documented in the project execution plan.

2. The project team comprises representative of the owner, project manager, engineer, consultants, and contractor right from the outset of all the phases of the project.

3. The individuals with correct construction knowledge and experience are employed at the planning stage.

4. The methods of construction are well thought of and taken into consideration before choosing the contractors for different activities.

5. The construction schedule is worked out and completion date predicted before the execution of the project.

6. The schedule prepared is construction sensitive.

7. Suitable measures are adopted during the planning stage to make field operations easy.

8. Recovery and recycling is discussed during the planning stage.

9. Considerations are taken at the planning stage to make site layout efficient for accessibility of contractor, material, and equipment to the required position on-site during construction, operation, and maintenance.

10. The contractors are not always awarded project based on the lowest bid, but other attributes such as their experience in the same type of work and quality of work are also given weightage.

9.3.2 Checklist for Design Development Stage

In this stage, the design team does analysis for alternate solutions and materials. The detailed drawing is finalized together with the major systems, materials, components, etc. All technical documents, specifications, schedules, and budgets are developed at this stage.

1. Construction schedule is discussed before the procurement schedule.

2. Proper surveys for site have been performed before the design process initiated.

3. Advance information technologies are used.

4. Design is reviewed by the contractor.

5. Elements of design are standardized, and technical specifications are simplified.

6. The concept of modular design and preassembly for project elements is discussed and implemented, if advisable.

7. Design has taken care of water conservation and site drainage systems.

8. Future flexibility has been taken care of, by simplifying and separating building systems and components, in due consideration for maintenance.

9. Alternate and environmentally safe building materials have been searched for, as options before finalizing the specifications.

10. Use of locally available building materials and methods is preferred.

11. Design and construction schedule has taken care of adverse weather conditions.

12. Due consideration is given to methods and materials that allow for the ease of renovation and deconstruction.

13. Recycled and reusable building materials are given preference.

9.3.3 Checklist for Field Operations Stage

Actual work on-site begins here, and it ends when the project is completed.

1. The site is under constant supervision.

2. The field task sequencing was done in order to minimize damages or rework, and scaffolding needs or congestion of constructor, material, and equipment on-site.

3. Temporary construction material/systems are used for efficient construction.

4. Introduction and promotion of new equipment/tools or modification in tools is done to save energy and time.

5. The contractors are given freedom for taking decisions on-site, regarding the use of temporary facilities or preassembly in case of adverse weather conditions.

6. Documentation work was done after the project or during the project, to preserve as feedback and use as lessons learned in the future projects.

7. The site responsibilities are clear and coordinated.

9.4 EXPECTED ACHIEVEMENTS

The implementation of checklist at various stages of the project shall account for tangible and intangible benefits. The incorporation of constructability principles is expected to ensure the following achievements:

1. The level of satisfaction will be high.

2. Feedback system is prepared for the future projects.

3. Due consideration was given to manage waste on-site and environmental pollution was reduced effectively.

4. The site management was efficient.

5. Better design was achieved because of involvement of different experts at the initial stage of design.

6. The owner is satisfied.

7. Project work was done as teamwork, and the professional bond can lead to working together in future project.

8. There was a significant reduction in the project cost.

9. The project was completed as scheduled without delays and disputes.

10. There was no communication problem among the stakeholders involved.

Index

Printed and bound by CPI Group (UK) Ltd, Croydon, CR0 4YY

22/10/2024

01777627-0002